Northern Bee Books
Scout Bottom Farm
Mytholmroyd
Hebden Bridge
HX7 5JS (UK)

www.northernbeebooks.co.uk
Tel: 01422 882751

ISBN: 978-1-912271-32-0

Cover design by sipat.co.uk
Printed by Lightning Source, UK

Mary Bumby

The first person to take honeybees to New Zealand

Anne Middleditch & David F. Bumby

Dedicated to the wives of the early Wesleyan Methodist Missionaries in New Zealand.
These courageous women devoted their lives to their husbands and the church and
raised their families in what were often extremely difficult and dangerous conditions.

Foreword

with acknowledgements

In April 2017 David Bumby, a visitor to the Chain Bridge Honey Farm visitor centre, noticed an error in the information about when, and by whom, honeybees were taken to New Zealand.

I had researched the information for the visitor centre over a period of a few years and, on a world map, had placed the dates when honeybees were taken to various countries around the world, and who had taken them there. As is stated in many text books I had written that the first honeybees in New Zealand were taken there by the Rev. Wm. Cotton in 1843.

David Bumby told us that it was in fact Mary Bumby, a forebear of his, who had taken honeybees to New Zealand in 1839, four years before Wm. Cotton.

I was keen to find out more about Mary Bumby and planned to write a short article about her for the visitor centre. When I started to research her story and learn about her family I realised that it was going to be a much bigger project and an incredibly fascinating story.

David had, over the years, built up an archive about Mary and her family which he kindly loaned to me.

The archive contained a great deal of important material including a transcript of Mary Bumby's journal which spanned the years 1838 to 1841.

I am very grateful to David for his help and advice - reading all the documents, suggesting improvements, correcting typing errors and grammar, for suggesting improved phrasing and for supplying so much useful information.

Special thanks to John Steele, author of 'Smales Trail', a biography of the Rev. Gideon Smales, (Mary Bumby's husband). 'Smales Trail' is a fascinating account of an incredible man who did so much to help the Maori at the time of the colonisation of New Zealand by the British Crown. The book was a great help when I was working on the chapter about Gideon Smales. John Steele very kindly read over that chapter, giving me very useful help and correcting some errors and generally being very supportive of the project.

Many thanks to Gillian Stephenson for doing the illustrations for the book.

I am extremely grateful to Jeremy Burbidge and Simon Paterson of Northern Bee Books for agreeing to take on this project and for publishing the book. I very much appreciate their help and advice with the editing and proofreading of the script, the design of the book and for their never-ending patience.

Anne Middleditch
2018

Chapters

The Family of Mary and John Bumby

In 1751 John Wesley, who with his brother Charles and John Whitfield, had founded the Wesleyan Methodist Church, was travelling and preaching in Yorkshire. When he stopped at Osmotherley William Hewgill and other villagers from Hawnby walked 12 miles to hear him. They were so impressed by Wesley's preaching they begged him to visit their village which he did the following year. A Methodist Society was formed in the village and the members continued to worship in the Wesleyan tradition. Mrs William Hewgill was extremely devout and was rewarded by the conversion of her daughter Mary when she was in her twentieth year

Mary's father had died when she was young and in 1800, the year after she joined the Methodist Society she lost her mother too. In 1803 she married John Bumby, a butcher who lived in Thirsk, Yorkshire and, in 1806, a baby girl was born whom they called Mary but sadly she died when she was just three years old. In 1808 their son, John, was born and two years later they had another daughter whom they also called Mary. Jane Elizabeth was born in 1813 but she died when she was just 16 years of age.

John and Mary's mother, who greatly influenced their lives, was keen for her children to be well educated, and encouraged John particularly to read and further his education. She instilled in her children her own unwavering faith in God and the Methodist Church and supported John in his decision to enter the ministry and become a preacher and devote his life to God and the Church.

John's father was hoping his only son would follow him into the family business but eventually had to accept that John had made up his mind to devote his life to the service of others and to train as a minister. He paid for John to study at a university in Leeds to help him further his education in the Methodist religion.

John trained as a minister and after some years travelling the country preaching he was selected to become a missionary and was appointed as Superintendent at the Mission House at Mangungu, New Zealand.

John's mother had died some years before he and Mary left for New Zealand but it must have been heart breaking for his father to bid farewell to them both knowing he would probably never see them again.

Mary Hewgill Bumby was born in 1775 and died in1831. John Bumby Senior was born in 1769 and died in 1848.

Mary Bumby

In 1839 Mary Bumby became the first person to take the European honeybee to New Zealand.

Mary was born in 1810 in Thirsk, Yorkshire, where she and her brother John and younger sister Jane Elizabeth lived with their parents who were devout Wesleyan Methodists.

In 1838 John Bumby, who had trained as a minister, was appointed as Superintendent of the Wesleyan Methodist Mission at Mangungu in Northland, New Zealand. Taking up his appointment entailed a five-month long voyage so Mary decided to accompany her brother as she was always concerned about his health and she could look after him and act as housekeeper. John loved eating honey so she decided to take two skeps of bees with her so that he could have some while they were in New Zealand.

Thirsk in the late 1800s.

Before joining the '*James*' which was to leave from Gravesend in September 1838 Mary and John spent some time visiting family and friends knowing they would probably never see them again.

Mary's Journal

Mary kept a journal of the voyage and in the first entry she described how heartbroken she was to leave her elderly father and the home where she had been so happy.

Thursday August 16th, 1838

"At two o'clock this afternoon I bid farewell to my aged father and much loved house, it may be, and I think it most likely, that I shall never again see the house of my childhood, and the spot where I have spent so many happy days. Oh how my heart bleeds at the thought that I may never see my dear father again in this world."

During the next weeks Mary and John bade farewell to friends in different parts of Yorkshire visiting Knaresborough and Boroughbridge and beyond then travelling to Liverpool where they spent the last few days before travelling to London where they arrived on 6th September.

London, September 16th

"Came to this place about 10 days ago with brother Mr and Mrs Hyde in order to embark for New Zealand. This morning saw the vessel in which we are to sail. Think it rather small, feel very dull with the prospect of a five month voyage in such a prison."

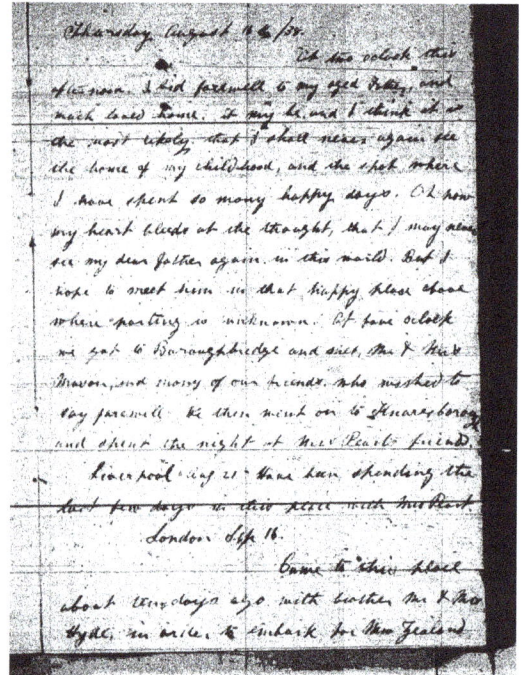

A facsimile of page one of Mary's journal.

There were to be 56 persons on board altogether so Mary was quite concerned about such a large company on a small sailing ship so the prospect of a long voyage must have been quite daunting.

On September 19th Mary and John arrived at Gravesend where they spent the day. It was raining heavily and after a walk around the town Mary described it as a *"very poor place"*.

On Thursday 20th September the travellers boarded the sailing ship *"James"* with a great many friends arriving to say goodbye and wish them well. One of Mary's close friends, Mrs Hyde, returned to London having travelled all the way down from Birmingham specially to see Mary once again which was a wonderful surprise for her. That same day they left Gravesend sad to be leaving their friends and anxious about the future.

After a rough passage as far as the Isle of Wight when many of the passengers suffered from seasickness, the ship weighed anchor and some of the men went over to the island to have breakfast and to buy a few last-minute items. One passenger had had enough and decided to go no further.

"The passengers all assembled on deck to see him depart and to give him three cheers. He left amid torrents of rain and causing a pig overboard, the pig saved."

Old photograph of Madeira.

By October 8th they were passing Madeira which Mary described as a beautiful island with lovely white-painted houses and she would have loved to have been able to visit – she wrote:-

"I would love to have some grapes as fruit is the only thing I can eat on account of the sickness".

Mary was troubled with seasickness throughout most of the voyage.

After passing close to the Canary Islands the ship crossed the Tropic of Cancer and the weather became very hot and oppressive which made sleeping in the small cabins very uncomfortable especially after a day in the scorching sun.

During the second half of October the weather was very calm so little headway was made and some of the ladies could go out for a pleasure trip in the ship's small boat which must have been a pleasant change. A shark was caught, part of which was cooked for breakfast but some of the passengers, Mary and John included, chose not to eat any wondering how some could eat

"such man-eaters no doubt that it had some bits of human body. They all appeared to be quite delighted with it and said it was very good".

On November 2nd the weather turned very stormy and Mary suffered terribly from the seasickness once again.

Some entertainment during the voyage could be had by watching the different fish, sharks and porpoises and even albatross. Mary also talked of seeing different little birds, and storm petrels which she called *"Mother Carey's Chickens".* Storm petrels evidently are looked upon by sailors as a bad omen and foretell bad weather. It was so hot that the pitch was coming out of the deck in streams. Mary wrote that:-

"cockroaches are in much abundance in my cabin that I could get very little sleep last night"

Following the trade winds the ship passed near Trinidad in the West Indies and Mary thought that the setting of the sun in that part of the world was the most beautiful sight she had ever seen.

On Sunday November 18th the ship crossed the equator and as is the custom "Neptune's visit" was celebrated by the sailors and passengers pouring water over each other.

"all the buckers and tubs in the vessel were in action, it was indeed a very strange affair. The ladies did not escape a good drenching as well. A subscription was made for the sailors".

Towards the end of November there was a serious case of food poisoning when 23 of the passengers and crew fell ill after eating fish that was either rotten or had been near rat poison. Mary's brother John was one of those stricken and she was extremely worried about him especially as there was no doctor or medical help on board. Mary described the meal of fish and potato as *"twice laid"* which presumably was left overs reheated.

There was a religious service on deck each morning for the passengers and crew and there are frequent references in Mary's journal who the minister on duty was and the text he used for his sermon and prayers. John was frequently the preacher and Mary always enjoyed attending his services. The missionaries' faith in God must have been a great comfort to them during such a difficult and frightening experience and would help them tolerate such awful discomfort.

One evening a lad of 16 years fell overboard but being an excellent swimmer he kept afloat until he could be saved by the sailors using the small boat which was lowered into the sea so it was a happy outcome.

It was still very hot when the ship passed the Tropic of Capricorn being 121° but soon the weather deteriorated and it became exceedingly cold and very rough, too stormy for anyone to go on deck.

The weather seemed very changeable with it being rough and stormy one day and fine and pleasant the next. Mary compared the good days to pleasant Autumn days in England.

Fishing was important as it provided fresh food for the passengers and porpoise and albatross were welcome additions to the menu. Mary described a large porpoise caught one day as follows

"the head and teeth of this fish are very beautiful, the flesh of which, when cooked, is very much like beef"

By the middle of December everyone was anxiously looking out for the Cape of Good Hope and on Tuesday 18th December the top of Table Mountain was seen like a thin band on the horizon. By 10o'clock in the evening the ship was safely anchored in Table Bay. As can be imagined the sight of land must have been wonderful.

South Africa – Table Bay – Wednesday 19th December

"Early this morning the Harbour master came on board to inquire the number of passengers and their names. At a much earlier time than usual the company all met for breakfast delighted with the idea of once again setting their feet on solid ground"

At nine o'clock the passengers went ashore and were met by one of the Cape Town missionaries who took them to his home and entertained them and looked after them during the day. Mary was delighted with Cape Town and described it as being

"a large well-built town with many very good streets and shops".

It was the Malay Christmas and Mary describes the town as being *"very gay"*. She goes on to describe the Malay customs – great feasts, taking food to the burial places of the departed and hanging lamps over the tombs. This must have been very interesting and enjoyable for Mary after the gruelling two months on the ship.

The following day 22 of their party had a trip into the countryside, travelling in two wagons pulled by 8 horses each. They travelled to a place called Wine Burgh, 10 miles distant from Cape Town, and there they were met by one of the local missionaries, Mr. Haddy, who entertained them and gave them a meal of cheese and bread which Mary enjoyed very much. They travelled on to the vineyards near to Constancia where they enjoyed the much-famed wine after which they returned to Mr. Haddy's residence where they had supper. They returned to Cape Town about 8 pm after what must have been a wonderful day.

After two days in Cape Town the party boarded the 'James' once again. That evening the weather turned stormy with very strong gales and the Captain was extremely worried in case the ship broke from her anchor like many of the other ships in the Bay. Later in the evening cries of *"lost, lost"* were heard by the crew and passengers and then they saw a boat drifting against the 'James' with five men and two women in it. The sailors soon saved them much to their relief. They were passengers on a ship bound for Sydney and had gone ashore in the small boat to get supplies so it was a happy ending for them but not so for passengers on other ships. One ship had lost 30 passengers since leaving England, another lost 9. In those far off days travelling by ship was a hazardous undertaking.

December 22nd

Once again the 'James' weighed anchor. The weather turned very cold which after the warm days in Cape Town must have been very disappointing. Christmas Eve and Christmas Day passed and Mary made a comment in her journal that it would soon be the end of another year. On the following day the Captain of an American Whaler with five of his crew came on board as they were in great rouble, being out of provisions. They had been at sea for 19 months, one man had been lost overboard and another had a broken leg so they were glad to receive help from the crew of the 'James' who provided them with potatoes, pumpkin, tobacco and grapes.

On New Year's Eve Mary Ann's thoughts turned to her family and friends, knowing they would be thinking of her and praying for her and John's safety.

1839

On New Year's Day Mary wrote

"Early this morning the gents announced the arrival of the new year by ringing the large bell. At nine o'clock our party met and held a prayer meeting. The weather is very rough almost breaking over our vessel in a most terrific manner. The water is in my cabin so that I cannot go to bed and I am so ill that I know not how to manage to keep myself up."

It was not a good start for the new year with the weather continuing to be very stormy. Mary described how the chicken coops, casks and everything moveable was falling in all directions with the sailors

"shouting in a most deafening manner".

The weather eased a little and the ship made good headway and on Saturday 12th January they came to St Paul's Island in the Indian Ocean.

"At ten o'clock this morning we got to St Pauls, an island in the Indian Ocean. It is about eight miles long and five broad. It is uninhabited except for wild goats and pigs. The Captain and several of the gents went on shore to fish. After an absence of five hours they returned with a great many which we had cooked for dinner and found them to be very good. The island is on South Lat 38°44' East Long 78°40'.

The fine warm weather did not last long and on the 16th it was very rough all day with all sail taken in and the dead lights put on. John had a bad fall in his cabin and injured his back quite badly and was confined to his bed for several days. Mary was feeling quite ill but managed to look after her brother.

Thursday 31st January

"At twelve o'clock today to our no small joy we saw land ahead of us about sixty miles off. All was bustle and delight at the thought of once again setting our feet upon solid ground after being at sea for nineteen weeks".

From an old drawing of Storm Bay, Tasmania.

The following day at one o'clock in the morning the '*James*' entered Storm Bay, Tasmania and at four in the afternoon sailed into the beautiful River Derwent. At 6pm a pilot came on-board but it was another 20 hours before they reached the harbour because of contrary winds.

Mary wrote

"We were met by Mr. Orton and Mr. Such. Mr. O. I think is rather a stiff brother."

As Mr. Orton hadn't made any arrangements for the passengers several returned on board to sleep (which must have been very disappointing after such a gruelling voyage).

On the Sunday, February 3rd, Mary accepted an invitation from a Mr. and Mrs. Heddlestone to stay with them during the time they would be in Hobart Town.

"I have a neat little room and the grapes are growing in great abundance round the window. I do not much like the look of my hostess, she is exceedingly cold and stiff in her manner, hope I shall like her better when we get a little more acquainted".

The following day Mary walked out to see the town.

"It is a large place with about twenty thousand inhabitants chiefly Europeans. The country is romantick and mountainous. The climate is very changeable – sometimes very hot and at others very cold. We have heard today that the Mission House in New Zealand is burnt down – the place where we are going".

This must have worried Mary and she would be wondering where they would live on arrival at the Mission.

(Note – the Mission House was burned down in August 1838 as described in a later chapter. This was before Mary and John had left England. It shows how difficult communication was at that time and how slowly news travelled. The Mission House was rebuilt by the time they arrived.)

During the five weeks spent in Hobart Mary and her friends enjoyed several outings. They met his Excellence the Governor Sir John Franklin who invited them to have dinner with him and Lady Franklin. Mary also visited a French frigate and went on board and another day her brother John took her to Langrough Park, a small place across the River Derwent where they had lunch.

Monday 25th February

"Went with a party of friends in a boat down the river about ten miles. We landed in the bush made a fire and had game tea and fish which we enjoyed very much".

Some outings were evidently more enjoyable than others. One evening the party dined at Government House with Sir John and Lady Franklin and they had a very sumptuous meal and wonderful entertainment. Another time Mary *"took tea and spent a very stiff evening at a Mr. Hacksons"*.

The last outing was very enjoyable as they spent the day in the country and had a picnic meal but Mary sprained her foot which made her lame for quite a few days.

On Friday 8th March all those, including Mary and John, who were bound for New Zealand boarded the '*James*' once more and the following day the ship weighed anchor and they were on their way. Just over a week later, after rough weather, the long-looked for coast of New Zealand was seen but because of contrary winds the '*James*' was unable to make land.

Monday 18th March

"At twelve o'clock we crossed the much dreaded bar of the Hokianga river. Captain asking the pilot to come on board to take us up the river a distance of about 25 miles. Mr. Woon who

lives a few miles from the Head, an amazingly stout man, came in a little boat and called out "is this the 'James'?" on being answered in the affirmative he was soon on board. I thought him a large specimen of the New Zealand brethren".

After being stuck on a sandbank the '*James*' managed to sail a few miles up-river casting the anchor for the night.

The following day, at 8am the '*James*' continued up river and by noon had arrived at the Mission Station at Mangungu and they were welcomed to New Zealand by Mr. Turner and Mr. Hobbs. Their arrival must have been a very strange experience. Mary wrote in her journal:

"Many canoes came to us crowded with natives who had pigs, potatoes, melons and Indian corn for sale. Their appearance is anything but pleasing to the eye of a stranger – they scarcely wear any clothing – an old coat or a blanket is generally all they possess".

On the following evening Mary and her companions met the great chief, Shamal Walker, whose Maori name was Nana. The ship's captain, to celebrate the safe arrival of the '*James*', fired many canon and had a display of fireworks which delighted the native Maori.

When Mary attended her first chapel service she was amazed and delighted to see such a large crowd taking part. She records that there were 450 Maori, all wearing blankets and seated on the floor, and 200 more outside as there was no room for them in the chapel.

The next entry in the journal is for Monday May 13th when Mary records that John and Mr. Hobbs along with 15 natives set off for the Bay of Islands to join the ship '*Hokianga*' to sail south to look for new mission stations. It was to be 11 weeks before Mary received a letter from her brother and by that time he was 200 miles from Mangungu. It was another three weeks before John returned and as can be imagined Mary was very happy to see him again.

Because of his position as Chairman of the whole mission in Northland, North Island, he was expected to travel widely in the region, preaching and also organising new mission stations.

The next entry in the journal is on Sunday December 15th which tells of John having departed once again, on board the ship '*Melrose*' bound for Launceston, Hobart Town and Sydney, Australia. When there he met up again with the Rev. John Waterhouse, who was the general superintendent of the Wesleyan Mission in Australia and the South Seas, and he met other missionaries as well.

1840

On New Year's Day, 1840 Mary once again wrote about another year gone and how she missed her family and friends back home. There was a ship leaving that day for England so Mary wrote letters to friends hoping they would arrive safely.

"On Tuesday 11th February Captain Hobson arrived here today along with five gentlemen and two church Missionaries, Mr. Clark and Mr. Taylor, he is appointed Governor of New Zealand."

(Note – Mary made no reference in her journal as to why Captain Hobson had come to Mangungu. In the chapter on the Waitangi Treaty it is explained that he had been sent out to New Zealand, as Governor, to organise the Treaty signing which would result in New Zealand becoming a British colony. Mary, in the absence of her brother, acted as hostess to the large number of people who arrived the following day so presumably had no time to write in her journal.)

The missionaries at Mangungu were expecting the arrival of the Wesleyan Mission ship, 'Triton' sometime in February. The ship had sailed from Bristol, England in September 1839 bringing a group of missionaries destined for Hobart, Fiji and Mangungu, but owing to bad weather and other setbacks it didn't arrive until May 8th.

On Tuesday 18th February John arrived back from Australia accompanied by Mr. Orton where he had been visiting Mr. Waterhouse and other missionaries. John had arranged to join the 'Triton' when it left Mangungu to sail south as he had work to do in the south of the region setting up new mission stations and continuing his work preaching and spreading the gospel. When the 'Triton' didn't arrive when expected the missionaries probably feared the worst but were rewarded for their patience when it finally arrived.

The next entry was Tuesday, May 8th when

"to our small joy the 'Triton' arrived. It got into the Hokianga last night".

Waiting for and looking out for the 'Triton' to arrive for five weeks must have been a worrying time and illustrates the difficulties and dangers of travelling by sea in those days.

The next ten days must have been very busy for Mary as a large party of missionaries, including Mr. Smales, arrived on the 'Triton' and she would act as hostess and would make sure they were comfortable and well fed.

By the 19th May preparations were being made for the 'Triton' to set sail and John was joining the ship to travel south to set up new mission stations but it was three days before the ship departed. Mary was very worried about her brother as his health was poor and she felt it was too much for him to undertake such a journey as he would have to travel back overland about 200 miles, once he had completed his work.

Monday June 1st

"Had a note Mr. S. offering me his hand and heart. Know not what to think of it as I do not want to make any engagement of this kind".

Because of her worry about her brother marriage was probably the last thing on her mind and she didn't accept his proposal.

An entry for June 9th described how the worry about her brother and his safety increased. Perhaps Mary had a premonition that all was not well with her brother as she had a very vivid dream about him.

"About 2 o'clock this morning I was awoke by the voice of my brother calling me by my name. I feel very unhappy about him as I am afraid all is not well with him. I thought I saw him standing by my room door but the moment I looked up he was gone. This appears very strange to me. I hope the Lord will be his keeper and that he will come back in safety".

The weather for the rest of the month was very cold and rainy which added to Mary's concerns about John. The bad weather had its disasters – a canoe with a wedding party on board overturned near the Mission House, all were saved fortunately but the wedding feast was lost.

A most distressing event happened on the 13th July. Mary received a letter of condolence from Mr. Taylor of the Church of England Mission

Mary wrote:-

1840 July 13th Monday

"About ten this morning I received a letter from Rev. R. Haydon Church Missionary consoling me on the sudden removal of my Dear Brother. I was in a moment, in this sudden way plunged into the utter most distress. A messenger dispatched immediately (to get information) *I was in the greatest anguish all day. I took but little food and any sleep departed from me."*

As can be imagined Mary was distraught, wondering what had happened to her brother but it wasn't until the next day that news arrived confirming there had been a tragedy and Mary and the others at Mangungu learned that John had been drowned when the canoe in which he was travelling overturned. Only six of the nineteen on board survived to tell the tragic story.

Mary was overcome with grief at losing her beloved brother. Her friends rallied round, trying to comfort her in her time of great sorrow.

(Note – the tragedy happened on 26th June and it wasn't until the 14th July that the sad news reached Mangungu which again illustrates how difficult communication was at that time and how long it took for news to get through. The tragedy is described more fully in the chapter on John Hewgill Bumby.)

It wasn't until the 16th July that Mary felt able to write letters to her father and friends back home, telling them of the tragedy. The following day Mr. Smales and Mr. Hobbs set off for the Bay of Islands hoping to find a vessel which would take them to the site of the tragedy 150 miles away. Their aim was to find John's remains which in a very small way would ease Mary's pain.

On 9th November Mary received a letter from Mr. Waterhouse saying he had just heard the tragic news of John's death and was overwhelmed with sadness at losing a good friend and wonderful person. Mr. Waterhouse was based in Australia so it took many months for the news to reach him.

Mary sometimes spent time in John's room at Mangungu thinking of him and looking through his papers. James Garland, one of John's companions on the stricken canoe, had survived and had managed to save the boxes of John's papers which he returned to Mary.

On Wednesday 9th December she wrote about reading over some of her mother's letters to John while he was in the Waltham Abbey Circuit long before his departure for New Zealand. This was in 1830 and a few months before her death. Her health by that time was not good and although she longed to visit him in the south she was not well enough to make the long journey. She wrote to him *"I now begin to think I shall never come to London or even see it, unless I look over the place where my lad is as I go up home."*

On Friday 18th December at eight in the evening Mr. Waterhouse with two companions arrived from the south on the 'Triton'. The ship got as far as the Heads in the afternoon then became stuck on the bar and the rudder was broken so the ship was in a very dangerous position. All were saved, fortunately, but it was a very sorrowful meeting for Mary as their thoughts were with John and his companions who had lost their lives

On Tuesday 22nd December Mr. Smales once again proposed marriage to Mary and this time she accepted.

"Well, I have made up my mind after much consideration and prayer to become a missionary's wife!! My soul is deeply humbled at the thought of the responsibility of such a situation. But I hope the Lord will help me by his grace to discharge the duties that will devolve upon me and help me to live to his glory."

There was a large party, about 20 in number, for Christmas Day and Mary felt the pressure of acting as hostess as she still felt quite ill.

Tuesday 29th December

"This morning I was married to Mr. Smales in the Chapel at Mangungu by Mr. Waterhouse. We have consented in the name of the Father and of the Son and of the Holy Ghost to bear each other's burdens and to become one forever."

1841

Over the next few days the visiting missionaries left for other parts and on the 28th January Mr. Smales and Mary also left Mangungu for Pakanae where Mr. Smales would continue his mission work in the region.

"We had a very rough passage of about eight hours. We had far too many things in the boat and only three hands. I thought more than once we should have been set up (uoset) but the Lord was better to us than our fears. I hope that our coming here will be for the glory of God and the good of the poor natives."

(Note - Pakanae is a small habitation on the Hokianga Harbour so would be 20 or more miles down-river from Mangungu. The Hokianga River, a major river, was quite navigable for sailing ships as both the '*James*' and the '*Triton*' were able to reach Mangungu and so it would be a hazardous journey for a small boat, laden with belongings, the crew and the Smales on such a big river.)

Probably due to having duties as a missionary's wife entries in Mary's journal are less frequent and quite brief but she does mention a visit by Lady Franklin on Monday 10th May with her attendants, a lady and gentleman, and three servants. Her arrival was an unexpected pleasure.

Friday 21st May was the first anniversary of the day John left Mangungu for the south and was the last time Mary saw him.

 "*He has been sleeping in the cold and greedy sea for one year. Oh how my heart bleeds at the thought.*"

Monday May 24th

The long looked for the '*James*' arrived with Captain Todd and as the wind was fine the ship went up the river to Mangungu. Mr. Smales, with the help of natives, got a boat and went out to the '*James*' and in the evening returned with a box from Mary's father. There were more packages for her on-board but some other cargo had to be taken out before they were retrieved.

(note – the '*James*', as did the '*Triton*', plied to and forth between New Zealand and Hobart and Australia bringing passengers, goods and letters that had arrived from England on larger sailing ships.)

October 21st Thursday

"*By the helping of God I was safely delivered of a son. Mr. S. and Mrs. Young were my only attendants. Mr. S. sent for a doctor who lives about ten miles off but he was two hours behind time. I had no nurse for the first three days except Mr. Smales and then had Mrs. F. White who spent a week with me*"

(Note – in her journal Mary always refers to her husband as Mr. Smales which to us may sound strange.   His first name was Gideon.)

January 20th 1842

"*This morning we received a letter from Mr. Stones. He sailed for England about a month ago. Sometime in March he will arrive at his long wished for home, the loved abode of health and peace.*"

Mr. Stones was a young man (whom Mary described as a "*very nice lad*") who had visited Mangungu on several occasions and had made quite an impression on everyone.

This is the final entry in Mary's journal as presumably being a wife and mother gave Mary little time for writing.

Friends of Mary describe her as a gentle, generous and thoughtful person who loved children. She was also an incredibly brave young woman who chose to accompany her brother on a gruelling five -month long voyage to an unknown country.

In her first few years of being a missionary's wife life was hard. Living conditions were very basic and she had to manage long periods on her own without Gideon when he was away travelling, carrying out his mission work.

For both Mary and Gideon their children were the centre of their world. Mary provided a happy home life binging up her children in the Christian faith and making sure they had good education.

Mary and her husband Gideon served the mission faithfully for many years in Pakanae and Aotea. There is more detail about Mary's life after her marriage in the chapter on Gideon Smales.

The Rev. John Hewgill Bumby

John Hewgill Bumby was born in Thirsk, Yorkshire, in 1808 to parents John and Mary Bumby. John's father was a butcher in Thirsk and he was their only son. He was a sensitive boy and had poor health but he showed a very determined and single-minded personality from a young age. He had an enquiring mind and enjoyed studying nature and history and was fascinated by the stories told by visiting Methodists who had travelled abroad.

John was very much influenced by his mother who encouraged him to read and extend his education and to devote his life to the Church and to the Wesleyan Methodist Faith. In 1823 when he was just 15 years old he preached to a gathering in the open air and also attended church meetings and discussed church matters with his elders. In that same year he converted to the Faith. He was described as an honest, frank and warm-hearted Methodist youth.

Although his father had hoped he would follow him into the family business he accepted the fact that John had made up his mind to train as a minister. When John was 18 years old his father paid for him to attend an Academy in Leeds to further his learning but it was not a happy time because of all the arguments and dissention among the Wesleyan Churches in Leeds at that time. He left after a few months.

John trained as a Methodist minister and in 1829 entered the Wesleyan Methodist Missionary Society and the following year commenced probation in Waltham Abbey. John became very well known as an excellent preacher and spent some years travelling throughout England spreading the gospel.

In 1838 John was chosen to go to the Wesleyan Methodist Mission at Mangungu in North Land, New Zealand, as Chairman, replacing the Rev. Turner. As mentioned earlier, John's sister Mary accompanied him to act as housekeeper and to look after him. Mary took two skeps of bees with her so that John could enjoy some honey which he loved.

Like Mary, John kept a journal of the voyage and of course recorded the same events such as seeing porpoises and albatross but he seemed to appreciate more the beauty of the sea, the way the light danced on the waves, the wonderful rainbows, the sunrises and the sunsets. He did however write about how homesick he was, missing his father and friends. He was concerned about Mary who was troubled by severe sea sickness for most of the voyage.

After the long voyage John and Mary arrived at Mangungu Mission House on Tuesday 19th March 1839 and John took up his position as Chairman in charge of the Wesleyan Methodist Mission in the whole area. For many months John travelled widely preaching to the Maori and Europeans, teaching them a better way of life and encouraging them to convert to Christianity and he also visited Australia to meet other Methodist Missionaries who were stationed there.

He was much loved and respected by everyone who had met him and he worked incredibly hard, travelling and preaching even though his health was not good. In a letter dated September 1839 he wrote about suffering from a very painful eye condition, ophthalmia, which was agonising and left him almost blind. His eyesight improved when the inflammation abated.

John learned the Maori language so that he could communicate with the natives more freely.

June 1840

On the 24th June, John preached at Wangaroa, the text being 'contentment' but having been away from Mangungu for many weeks he was now anxious to return. While passing through Waikato district (in the region of Auckland) he called on the Rev. Fairbairn of the Church of England Mission to ask advice on the best route back to Hokianga. The route Rev. Fairbairn suggested involved a 70-mile trek after crossing a wide estuary, the Heads of Kaipara, but John was not happy about this as it would be a very strenuous journey and he was foot sore and weary.

John preferred to go by sea as it would be less difficult and he consulted with several Maori who suggested the best sea route. Several natives were to accompany John on his journey back to Mangungu and so there would be about twenty altogether. Unable to hire a boat – Mr Fairbairn was reluctant to lend his boat as he didn't trust the natives to look after it – John decided to travel by canoe.

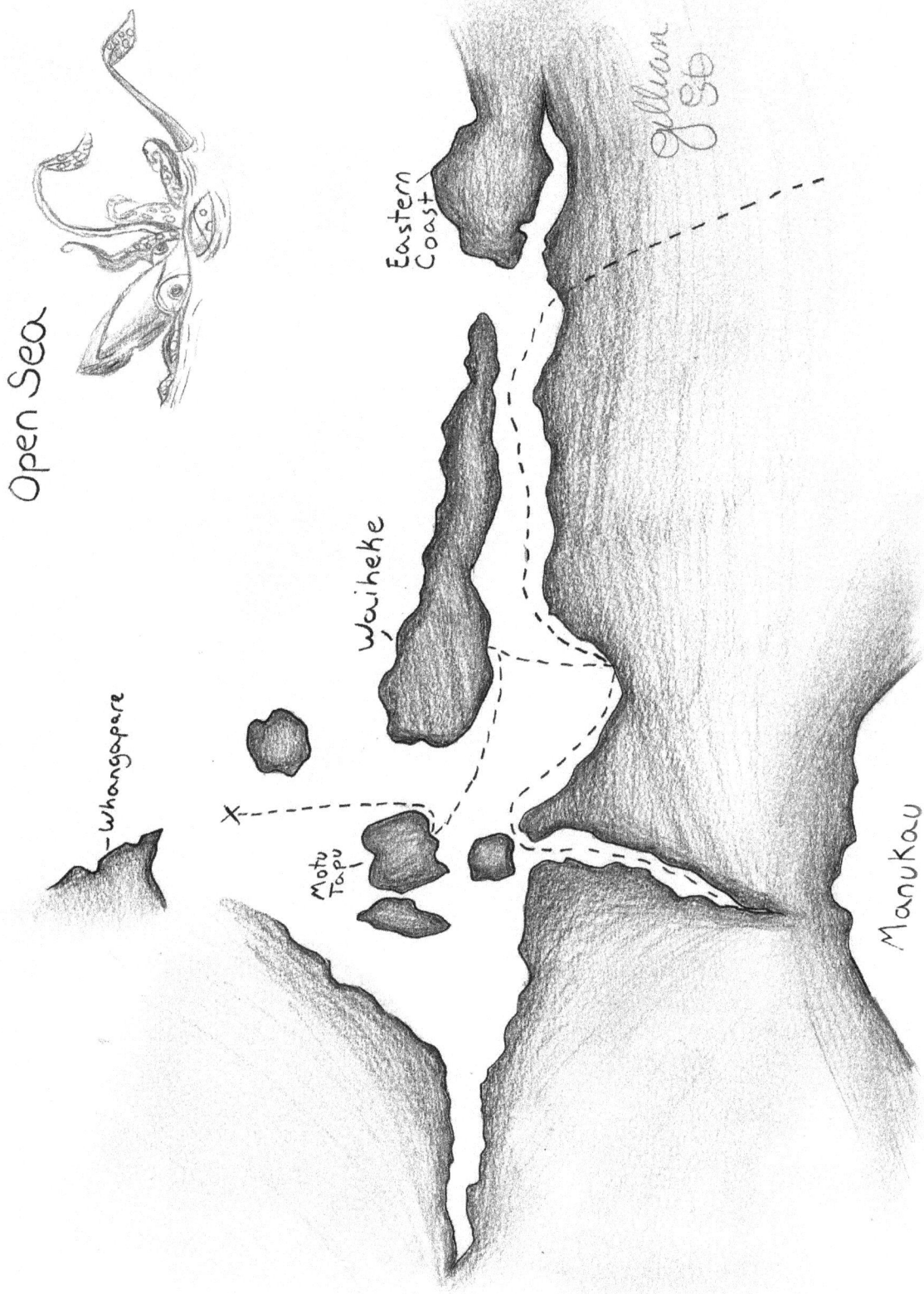

Open Sea

Eastern Coast

Gulliun Sd

Waiheke

Whangapare

X

Motu Tapu

Manukau

As was the custom at that time some of the Maori adopted English names if they were attached to an English community, like the Mission, and very often the names had a religious connection. Among the Maori accompanying John were Morley, John Beecham, Theophilus, Nehemiah and a lad from Tonga called Joel. James Garland whose Maori name was Hemi Kamara was also with the group.

On the 24th June John and his companions travelled to the Island of Waikeke where Morley had family. After staying there overnight they set off the following morning in a canoe supplied by Morley's family. From his residence Mr Fairbairn watched their leaving and he was full of foreboding.

The weather was calm and that day they reached the small island of Matu Tapu where they stayed overnight. The following day, the 26th June, John was so eager to set off that the Maori had very little time to cook and eat their food so were not very happy at being rushed and some looked on this as a bad omen.

The helmsman was called HuHu and it took him all his skill to keep the canoe right even in calm weather. About midday it was decided to hoist the sail to catch the breeze. One man stood to raise the sail but others, thinking they could help, stood and moved to the same side and with all the weight on the one side of an already top- heavy canoe it overturned and all were flung into the sea The Maori were good swimmers and soon righted the canoe and managed to get John back on board. The canoe again overturned as those still in the water tried to get on board and once again they were all thrown into the cold sea. The natives again saved John, not heeding their own safety, and they managed to get him onto the upturned hull where he was held safely by James Garland. By this time John was in a very poor way and was heard to mutter "*Ka matu Ka matu this is death*". A huge wave washed over the canoe once more and again John was swept into the sea and he sank, never to be seen again. Joel, who had tried to save him, was heard to cry out "*He's gone. The white man has gone*" then he sank too.

Some of the natives clung to the sail but it sank taking them with it. Of the 20 on board only 6 survived to make it back to land to tell the tragic story. They had managed to right the canoe and to save some of John's boxes of papers.

Among the survivors was James Garland who drew a picture of the tragedy as he would have found it difficult to write an account in English. The following sketch is based on his drawing and illustrates clearly the very rough seas and the bodies of the Maori and John Bumby being thrown into the water with such tragic consequences.

Communication, either by letter or word of mouth, was very difficult at that time especially in such a remote area so it was many weeks, even months, before all of John's colleagues heard about his death.

It took James Garland and his companions nearly two weeks to reach Mangungu to tell of the tragic event so no one there knew what had happened to John. By a tragic twist the

Drawing depicting the canoe tragedy.

Rev Richard Taylor of the Church of England Mission at Waimate who had heard the sad news wrote a letter of condolence to Mary which she received before the news had reached Mangungu. As can be imagined Mary was overcome with worry wondering what had happened to her brother and it wasn't until the following day that her fears were realised when the sad news was confirmed.

Everyone who had known John Bumby were overwhelmed with grief at losing such a wonderful person and also the mainstay of the whole Mission. It took many months for the missionaries to come to terms with the situation and to continue their work.

Commemoration

John Hewgill Bumby is still remembered and on the 150th anniversary of his death a special memorial event took place on the 26th June, 1990. A record of his life and his work for the Wesleyan Mission, as well as a few of his many letters, was published as part of the commemoration.

Rev. John Bumby,
Sesquincentennial
Commemoration,
June 26th, 1990

Some of those who had attended the ceremony sailed out to the approximate spot where John Bmby and his companions had drowned. A short service was held with a Bible reading and prayers and sprigs of rosemary and a wreath of Karaka leaves were strewn on the water. It was a very moving experience for all who were there.

IN MEMORY
OF THE
REV. JOHN H. BUMBY.

The plaque in memory of the
Rev. John Hewgill Bumby.

High Street Methodist Chapel, Auckland where the
plaque was first placed. It was later moved to Pitt Street
Church when the chapel was no longer in use.

The Rev. Gideon John Smales

The Rev Gideon Smales.

The Smales of the shipbuilding industry was one of Whitby's most prominent families in the Victorian era when Whitby (Yorkshire) was the third largest centre of shipbuilding after London and Newcastle.

The Smales Shipbuilding Business

18th C. - Smales Shipbuilding business established

Early 19th C. - Gideon Smales 1st took over the business

1817 – his son Gideon 2nd took over the business on death of his father

Eventually – his legitimate son Gideon 3rd took over the business

The Smales family business in the 18th century was involved in the shipbuilding industry by manufacturing marine supplies, cordage and block and tackle and it is said that they made numerous items for the famed vessel, the 'Endeavour,' Captain Cook's ship.

When Gideon 1st took over at the start of the 19th century he expanded into shipbuilding and built up a thriving business. Sadly in 1817 he died suddenly at just 51 years old and his son, Gideon 2nd, in later years known as 'Owd' Gideon, took over when he himself was just 22 years old and over the years he continued to expand the successful business.

Two years earlier he had begun an affair with a local lady called Susannah Nares who was a few years older than he was. They had a son, Thomas Smales Nares and two years later, on 26th. October 1817 a second son was born and they named him Gideon John Smales – the Gideon of this story. A third baby boy was born in 1820 and he was named James Nares so it is doubtful if James was Gideon 2nd's son as he wasn't given the surname 'Smales'.

Soon after James' birth Gideon 2nd parted from Susannah and a year later married Maria Wakefield, a shipbuilder's daughter, and the following year they had a son, Gideon 3rd. followed by 13 more children over the next 23 years.

Gideon (of this story) lived with his mother and his two brothers and, after Susannah married, also his stepfather David Gaskin. It is believed that his natural father however continued to support him and his brothers, both financially and emotionally, and made sure they had a good education. Gideon was very much part of the large extended family and was brought up to have strong Christian and family values. His stepfather's family was closely connected to the Wesleyan Church and this may have influenced Gideon into deciding to enter the ministry and to devote his life to the church and to God.

For the first 20 years of his life Gideon lived happily in Whitby with his family and had a good relationship with his natural father.

Gideon would realise that his younger half-brother, Gideon 3rd, would inherit the shipbuilding business at some point so this may also have influenced him to follow a different path. Being just an ordinary employee in the business wouldn't appeal so being a committed Christian and having been brought up in the Wesleyan faith Gideon trained as a minister. When he was 22 years old he was ordained at the Liverpool Conference of 1839 and was selected to become a missionary. In that same year on Saturday 14th September he left England to join the Wesleyan Mission at Mangungu in Northland, New Zealand. He sailed with several other missionaries on the brig 'Triton' and after a hazardous voyage arrived safely at Mangungu on the 7th May, 1840.

On the 1st June, a few weeks after arriving at Mangungu, Gideon proposed marriage to Mary Bumby by letter. She didn't accept his proposal mainly because she was, at that time, very worried about her brother John's health and his safety as he was many miles away travelling and preaching south of Mangungu and she didn't feel ready to become a missionary's wife.

On June 26th John Bumby was tragically drowned and it was several days after the tragedy before the sad news reached Mangungu. As John Bumby had been the superintendent at the Mission the next in command, the Rev. John Hobbs, became the principal missionary at Mangungu and the wider region.

Gideon and John Hobbs, unfortunately, did not have a good relationship from the start of their association. They, however, travelled together on the day following the news of John's death, by boat, the 150 miles to the site of the tragedy to try to find John's remains. This would in some way help Mary come to terms with her grief. Brave attempts had already been made to try to find John's body but eventually it was realised that he would never be found.

For the next few months Gideon travelled around the region, and further afield, carrying out his missionary work among the native Maori, preaching and spreading the gospel.

In December 1840 Gideon once again proposed marriage to Mary and this time she accepted his proposal and they were married on 29th December at Mangungu, the ceremony being carried out by the Rev. John Waterhouse. Mary was still living in the Mission House as there was no other suitable accommodation for her and after her marriage to Gideon he, naturally, moved there too making use of John Bumby's personal belongings and furniture (as husbands did in those days).

This made the Rev. Hobbs furious, mainly because he was living in Pakanae, a few miles down- river with his wife, Jane, and he felt as chief superintendent they should be living in the Mission House. As he had designed and built the Mission House after the previous one had been destroyed by fire in 1838 he felt he had a claim to it. Eventually it was decided Gideon and Mary would move to Pakanae and exchange houses with the Hobbs on the same day. This, unfortunately did not work out as planned and owing to bad weather Gideon and Mary were still in the Mission House when Mr. and Mrs. Hobbs arrived to take over. The situation did not improve relations between the two men, especially when Hobbs insisted Gideon and Mary move out into a small cottage nearby until they could move down to Pakanae.

Gideon and Mary's little house at Pakanae.≠

Another awkward situation arose after Gideon and Mary had moved to their new home. Packages of letters were still arriving from England addressed to John Bumby and these were given to Mary as his next of kin. The letters were mainly official documents from the Wesleyan Society headquarters and intended for the superintendent but also there was private mail for Mary. Hobbs insisted that he should be given the unopened packages and he would pass on any private letters to her but Gideon wouldn't agree saying she had the right to receive and open the packages herself.

Hobbs had fallen out with other missionaries so Gideon had support and friendship from some of his colleagues but even so he found the strain of the situation stressful and he was very keen to move south to the Mission Stations in the area of Port Nicholson (later called Wellington) where missionaries were needed. It would, however, be many months before he would achieve this. Even although Hobbs had a poor opinion of Gideon saying he was merely a novice and only capable of doing bookbinding and office work (Gideon had learned bookbinding while still in England) he wouldn't release him until a replacement was found. He accused Gideon of not working hard enough in the Newark mission field around Pakanae and that he hadn't made much effort to learn the Maori language. Hobbs also felt Gideon spent too much time helping Mary with her gardening and her bees, time that would be better spent going round his circuit and carrying out his duties converting the local Maori.

The birth of their baby son, John Bumby Smales, on 21st October, 1841 brought great happiness to Gideon and Mary and the following year, on 14th December 1842, a second son, Horatio Hewgill Smales, was born which added to their happiness.

Gideon's application to move south stalled because a replacement still hadn't been found but fortunately the Rev. John Whitely, who became a good friend to Gideon and was stationed at Port Nicholson, recommended that he be relocated to the south, and his destination was to be Porirua, a small rather isolated area just north of Port Nicholson.

Gideon started to make plans for the move and, despite being advised to wait for the return of the 'Triton' he decided to take the first ship going south which was the 'Guide', an old brig which was hardly safe or comfortable. Gideon, Mary, little John and baby Horatio set sail on 7th February 1843 and after a very stormy and difficult voyage arrived at Port Nelson on the north coast of South Island, just across Cook Strait from Port Nicholson. Little Horatio, just a few weeks old, caught a chill because of the cold and damp on board the 'Guide' which was of great concern to his parents.

Gideon and his family remained in Port Nelson for some weeks where he helped his colleague, John Aldred, set up the mission there. While there he was becoming more concerned about the welfare of the local Maori and felt that they were being exploited by the New Zealand Company. The company was buying up huge tracts of land from the Maori at ridiculously low prices and selling it on in small farm-sized parcels to the future settlers who were planning to emigrate from England. The Company was making huge profits on the deals.

Gideon was sure trouble was brewing between the Maori and the prospectors because of the land deals so he decided it was time he made his way to Porirua which was a stronghold for the Maori tribe Ngati Toa led by their warlike chiefs Te Rauparaha and Te Rangihaerata. Te Raupahara was extremely angry over, particularly, one block of land which the Europeans wanted but which he was unwilling to part with. Because of his good relationship generally with the Maori and his fluency in the language Gideon felt he could perhaps do something to alleviate the situation.

He knew that the living conditions would be very basic in Porirua so decided it would be best for Mary and the two little boys to remain in Nelson until he felt it was safe for them to join him particularly as little Horatio was still a very sickly baby. He managed to get a passage on a ship, the 'Hokianga' which would take him to Porirua harbour.

He found a suitable spot there to build a little hut for himself then commenced his mission work in Porirua. Part of his duties was also to spend some time back in Port Nicholson, helping John Aldred to set up the mission there and in mid - June, while there, the trouble between the Maori and the Europeans escalated into violence and tragedy. A Government brig sailed into Port Nicholson harbour with some wounded men, European and Maori, with news of a massacre that had taken place at Wairau valley near Port Nelson. There had been a confrontation between the two Maori chiefs and the New Zealand Company officials, each claiming ownership of the prized Wairau land. When the Government police tried to arrest and handcuff the Maori chiefs violence broke out resulting in 20 English and 4 Maori being killed. What made it more gruesome was that the Maori had taken the Eng-

Representation of a Maori Chief in New Zealand in the 1800s.

lish prisoners and then proceeded to shoot or murder them with tomahawks in revenge for one of their women being killed.

Many times in his life Gideon had shown tremendous courage in the face of danger and when the Maori chief Te Rangihaeta arrived back at Porirua after the massacre it was one such time. As their canoes landed Gideon went down to watch their arrival. He sensed trouble when confronted by the chief who was wielding a tomahawk and was ranting and raving in a most violent manner against the missionary, threatening to kill him. Gideon stood his ground and kept his nerve even when the chief approached threatening to cut off his head. Speaking in Maori Gideon tried to reason but it was only the intervention of the other Maori that prevented him being killed. Gradually the chief calmed down and slunk off and after what must have been a terrifying experience Gideon collected his belongings and returned to his small raupo hut. ('Raupo' is similar to our bulrush.)

The incident is referred to as the Wairau Affray or Massacre and is described more fully in the chapter "The New Zealand Company and the Wairau Affray".

About this time the Society decided the mission at Aotea, a rural area beside a west coast harbour south of Auckland, needed to be revitalised and offered the position of mission-ary in charge to Gideon. They knew he was popular among the Maori, was fluent in their language and was young and enthusiastic. He would be taking over from the Rev. Turton.

Gideon was only too happy to accept as he was keen to move away from the Port Nicholson area with the constant worry of trouble, the very basic living conditions and the poor land and cold climate. Aotea was in a very fertile area with a warm climate which would be better for Mary and the boys and they could establish a proper family home.

While Gideon had been at Porirua his brother, James Nares, travelled to New Zealand to visit him. James, sadly, could not find any information about his whereabouts so had to return to Whitby very disappointed. Porirua was a very isolated area out in the 'bush' with poor communication but Gideon would meet up with James and the rest of his family when he visited England.

In November 1843 preparations for the move were made and with their belongings packed they were ready to sail at the beginning of December. Sadly, on December 1st, little Hor-atio, not quite 12 months old and who had never recovered from his illness, died in his mother's arms. Desperately sad but with no choice but to board the ship Gideon and Mary wrapped up the little baby and took him with them. After another rough voyage they arrived at Kawhia, near Aotea, and were greeted by their friend John Whitely who arranged a funeral for the little boy before they travelled on to Aotea.

Gideon and Mary settled into their new home and despite their sadness they looked for-ward to a better life.

Gideon Smales at Aotea

The coastal region around Aotea was very fertile and with a mild climate was perfect for growing crops. Gideon and Mary found that potatoes and fruit such as apples, peaches, cherries and grapes were already being grown and there was an abundant supply of fish and sea- food so it was a much healthier and pleasant place for them to live in.

When Gideon took over from Mr. Turton he was still only 26 but he had ambitious plans for the Mission and worked tirelessly to accomplish them all.

Gideon's first task was to raise funds to pay for the building of the Mission House and he aimed to raise the sum of £120 in as short a time as possible. Donations of £1 each from local native Maori and £10 from Gideon himself and three of his friends soon realised the required amount for him to start his building programme. It took just two years for Gideon to complete the build of the Mission House which he called 'Beechamdale', named after John Beecham, one of the secretaries of the Wesleyan Mission Society in England who was always supportive of his work.

Gideon, being well educated himself, was very keen to promote education at Aotea so his next project was to devise an education plan for the local native Maori to follow. His syllabus was centred on mathematics, English and the scriptures. An exam system was introduced in 1845 and the exam results showed that some of the native Maori could repeat from memory the first four chapters of one of St Paul's Epistles.

In 1847, after three years in Aotea, he sent a report to the Wesleyan Society in England detailing the good results achieved, giving the number of Maori baptised and converted and the results of the English and arithmetic exams.

As well as promoting education Gideon was keen to establish farming at the mission. When he took over there were no farm animals at Aotea apart from a few wild goats which provided milk and meat. Gideon decided that having more animals would benefit the farming side of the mission and would provide food for the community. Within two years he had purchased 21 sheep, 6 horses, 12 cattle and of course there were still several goats.

Within his first year Gideon had planted 30 acres of wheat and in the next two years the acreage had increased to 600 acres, shared between the mission and some independent Maori tribes. When the wheat was harvested it was bagged up and shipped to Auckland for milling. Some of the flour was sold on to help defray expenses at the mission and some was used for bread making for the mission and the local Maori. Oats, barley, potatoes and fruit were also grown.

Gideon decided that having his own flour mill would cut out the effort and expense of transporting the grain to Auckland and would increase the efficiency of the farm. He contacted Stewart Mc Mullen, a millwright, to oversee the building of his mill which was the first water-powered flourmill to be built in New Zealand.

All this not only improved the diet of the Maori but provided them with an income, raised morale and gave them a sense of purpose and encouraged a more civilised way of living.

As Gideon expanded his farming project, improved the crop growing and invested in more animals he was able to make a profit by selling surplus products. This, he felt, would help to eke out his meagre salary and pay for necessities for the mission, thus saving the Wesleyan Society money. The officials in England, unfortunately, did not see things that way. They were very critical of Gideon and accused him of profiteering for his own benefit. Other missionaries were raising funds through farming and they too came under scrutiny.

One problem was that as their work in New Zealand expanded the missionaries needed more funding to cover mounting expenses. The Wesleyan Missionary Society at that time was experiencing financial difficulties but still didn't appreciate that Gideon and others were trying to help the situation by raising funds through farming. There was even discussion about making the New Zealand Mission independent and leaving them to solve their own problems.

There was a great deal of unrest generally in the areas around Port Nicholson and Mangungu in the mid 1840s with the threat of violence and attack on Europeans from some Maori tribes because of land ownership. Many settlers found that life in the new country was not as it had been portrayed, the land they had bought was useless and many of them sold up and returned to England. Another problem was that the Europeans sold or bartered alcohol, guns and other weapons with the Maori which just created more problems. The rapid development of towns like Port Nelson and Port Nicholson before the structure of those townships was organised resulted in chaos and unrest.

Despite all those problems Gideon and Mary enjoyed their first few years at Aotea, the happiest years they had experienced in New Zealand. Gideon worked extremely hard carrying out his mission work and expanding the farming and Mary had her hands full with her growing family and her garden and bees. They both submerged themselves in their work and their growing family. Their third child, Mary Anna, called Polly, was born on 6th. September 1844 and three years later Susannah Jane, (Rosie) named after Gideon's mother, was born on 12th. May 1847. While still at Aotea Gideon and Mary had three more children, Gideon Hewgill born in 1848, Felicia Clementina born two years later and Sophia Elizabeth in 1852. The children were the centre of their world

When Gideon was at Aotea and thinking of the future life for his family he decided to invest in land as a sort of insurance against what could be an uncertain future In 1851 the Government had settled some land claims involving original Maori land and the result was there were some acres available for sale to resident Europeans. Gideon took the opportunity and initially bought 127 acres at £1 per acre in the East Tamaki area near Auckland and the following year purchased a further 413 acres. This land, his farm which he called Hampton Park, would become his retirement home.

Questions were asked as to how Gideon could afford to buy land and he was again accused of using mission funds for his own gain. This was completely unfair and untrue as Gideon did have his own means to buy the land. Mary would have money or property from her brother's estate and Gideon had property in Whitby and he also obtained loans to cover the cost of the extra land.

The continual criticism by the Wesleyan Society that Gideon and others endured persisted for years and was one reason that Gideon felt that after years of incredibly hard work on behalf of the mission and the local Maori it was time to move on. Gideon had made two proposals to the Superintendent in Auckland – to have a year's sabbatical to recharge his batteries as his health and that of his family was not good or to relocate to the town of Auckland. Two of his children were at school there and life in the town would be easier for the family. Both proposals were turned down and the result of the disagreements and accusations levelled at Gideon and because he then refused to be relocated to another mission in Australia, he was suspended from the NZ Wesleyan Mission. In 1856 after 12 years of extremely hard work at Aotea, and still only 38 years of age, Gideon decided he had endured enough controversy and made the decision to move with his family to his farm at East Tamaki where he remained for the rest of his life.

Gideon did not completely cut himself off from the church but continued to take church services for the next 20 to 25 years, preaching in his own church and conducting services for other ministers in the area. He also retained his membership of the British branch of the Wesleyan Missionary society.

Life at Hampton Park

Gideon had ambitious plans for Hampton Park. His first challenge was to make the land productive and with the help of local Maori workers he cultivated 100 acres, planted crops and harvested them in his first year. During each successive year for three years he improved and cultivated a further 100 acres and turned the rather poor land into fertile farmland.

He established a milking herd, invested in horses, built sheds and animal housing and with his usual determination and sheer hard work he established one of the best farms in the area. Other major work carried out was the erection of fences, building of stone walls and construction of a road which connected the farm with the main highway to Auckland – at that time, via the sea. This project was so important for the transporting of crops and goods to Auckland and neighbouring farmers were very grateful for this as it was a benefit to them as well.

He had been very fortunate to meet James Stewart, an accomplished builder, who not only built the stone walls (some of which still stand to this day) but also did a great amount of building work on the farm.

Having lived for the last 15 years in very basic houses Gideon was determined to build a proper family home for Mary and the family and his vision was a typical English stately home surrounded by beautiful gardens. Until he achieved this they lived in a small cottage on the farm.

Once the farm was well established and was becoming a very profitable business Gideon and Mary's thought turned to the further education of their children. They had realised quite early on that young John was a very bright little boy with great potential. From a young age he showed talent as an artist and was an able student. While still very young he would sit for hours at his little desk copying all the illustrations in the London Natural History magazines and writing out correctly their scientific names. In 1855 he painted a very good water colour of the Mission House and other buildings at Aotea where the family had moved to in 1843.

(Note - The painting is still in the possession of the Smales family and is the only record of the Mission House and buildings as they were eventually demolished.)

Mary taught John at home for a few years until he was old enough to be enrolled at the Wesleyan College in Auckland where he remained until he was 18. Gideon knew that any further education in New Zealand, just emerging as a British colony, would not be adequate for John's future career, whichever path it might take. Within both the Bumby and Smales families education was of supreme importance.

The decision was taken that John, accompanied by his father, would travel to England and there Gideon would enrol him at the Wesleyan College in Sheffield, from where he would go to university. In February 1859 Gideon and John boarded the 'Sussex' in Auckland for the long voyage back to the homeland. Gideon didn't intend to stay long on this visit as he and Mary had already planned to return to England with young Gideon and the girls in two years' time in order to enrol them in schools as well. It was nearly 20 years since Gideon had set sail for New Zealand and it was with great happiness that he was able to see once again his mother, Susannah, and his brothers and his father and extended family in Whitby. After enrolling John at the college in Sheffield where he would be under the tutelage of Dr. Waddy. Gideon returned to Hampton Park, arriving in December of that same year.

As well as his farming work Gideon was still preaching and continuing with his church work but he was concerned that there was no proper place for worship so his next project was to build a little church on his land at Hampton Park. He drew up plans and James Stewart and his workers set to, quarrying the hard basalt stone from the quarry on the farm, and completed the build within two years.

Tragedy was never far from the lives of Gideon and Mary and in the autumn of 1860 their young son, Gideon aged 12, was thrown from his horse while riding around the farm. He landed heavily against a tree stump and suffered terrible internal injuries from which he never recovered. His parents were inconsolable, having already lost a son, and their only other son was back home in England.

Little Gideon was buried in the family vault of the still unfinished church.

James Stewart was determined to have the building of the church finished as soon as possible and this was achieved by November 1861. On Sunday, 12th January, 1862 the church

St John's Church, Hampton Park, East Tamaki.

St. John's was officially opened with church services, ceremonies and soirees to which all the local people came and there was a great feeling of camaraderie and enjoyment. Gideon announced that it was always his intention that the church would be used by all denominations and it is still in use today as a place of worship.

The visit to England

Well before the completion of the church Gideon and Mary had been making plans for their extended visit to England. They were to remain for a year or two so that they could settle the girls in their schools and spend time with family and friends and see more of John.

Over the years Gideon had continued to increase the number of animals on his farm and rather than concentrating on his milking herd he became interested in breeding sheep and eventually built up a herd of 800 but still retained a few bullocks, heifers, horses and milking cows. Part of Gideon's preparation for the extended visit was to sell all his farm stock, his machines, equipment, ploughs, harrows, furniture, books – in fact everything apart from the homestead and farmland. All was sold at a huge auction sale with no regrets as Gideon's philosophy was that it was all mere goods and chattels and could be replaced when he returned.

Soon after the ceremonies for the opening of the church Gideon, Mary and the girls excitedly boarded the ship, the '*Northumberland*' and left Auckland on Saturday, 18th January 1862.

After a calm start to the voyage the weather soon turned very stormy and for weeks all the passengers and even some of the crew suffered very badly. Mary, whose health was not good anyway, was very ill and was suffering from consumption as well and on March 22nd, just about half way to England, she passed away. Gideon and the girls were grief stricken and two days later Gideon read the funeral service and committed his wife's body to the sea.

It was a sad meeting with John when they sailed into London in May after four gruelling months at sea. Gideon and his daughters, Mary Anna (Polly) (18), Susannah (Rosie) (15), Felicia (12) and Sophia (10) then travelled to Whitby to see Gideon's mother Susannah and the rest of the family.

Gideon met up with his father, 'Owd' Gideon, and his half- brothers, Gideon 3rd. and Charles who were now running the shipbuilding business very successfully. His stepfather, David Gaskin had died three years previously and Susannah was now 72 years old. Once the girls were settled in school Gideon looked for something to occupy himself and decided to turn to writing. Although he fully intended to return one day to his farm in New Zealand he decided to remain in Whitby for some time, especially as news coming out of New Zealand was very worrying. The troubles over the land deals had escalated into warfare with brutal killings and murder. He was very concerned about the welfare of his Maori friends and what the eventual outcome would be.

As it turned out Gideon remained in England for five years, spending most of his time at his old home in Esk Terrace in Whitby. He was keen to see the girls happy into their schools and of course would see more of John who was still at Sheffield College and would eventually be accepted into Cambridge University.

Gideon was not one to remain idle for long and he soon started on his project which was to write a book chronicling all the publications of books and other works by Whitby residents. This took him five years and involved a huge amount of research and reading of records, documents and letters. The completed book "Whitby Authors and Their Publications with the Titles of all the Books Printed in Whitby AD 670 to AD 1867" was published by the Whitby publishers J. Horne and Sons in 1867. This was a tremendous achievement and the book is now regarded as a very important work of social history.

When Gideon, still a relatively young man of 47, had been widowed for three years he decided it was time to think about remarrying now that the accepted time for mourning was over. It would be good for the girls to have a mother figure in their lives as well as companionship for himself. He was introduced to a young woman, Maryann Baxter, the daughter of ship owner James Baxter. After a short courtship, Gideon and Maryann were married on November 30th. 1864.

Sadness again entered Gideon's life. His mother, Susannah, died just a few weeks after his marriage to Maryann and John became ill with TB and had to leave his studies in Cambridge and return home to Esk Terrace. This was a dreadful blow to Gideon and the girls. John's health did improve over time and he was eventually able to resume his studies at Cambridge.

There was however some happiness to look forward to for Gideon. Maryann became pregnant and their little son, Arthur Baxter Smales, was born on 2nd. September 1865 and his eldest daughter, Polly, became engaged to Samuel Chadwick, a young man from Malton in Yorkshire. Samuel had spent three years in New Zealand as an expert geologist and after they were married they returned there which pleased Gideon very much.

When little Arthur was one year old Maryann became pregnant again and their second son, James, was born in April 1867 but survived only a few days.

Once Gideon had finished his book and it had been published he began to think about returning to New Zealand. One problem was that James Baxter had made it clear to Gideon that his daughter Maryann would never leave England to live in New Zealand. She was a rather frail young woman and he felt she would not survive the hard life there or even the long voyage especially with a young child.

For Gideon the time seemed right for his return home. News coming from New Zealand was much better regarding the land wars and the trouble between the Maori tribes and the settlers had virtually ceased and everything was much calmer and peaceful.

Before Gideon had left for England he had arranged a lease on his farm so he knew the land would be well cared for and the fences and buildings well maintained but he was very keen to return and carry out all the plans he had for developing Hampton Park.

He also received very pleasing reports about his little church St. Johns. It was being used extensively for church services, weddings, christenings, funerals and social meetings by all denominations and had become the centre for the local community.

As Gideon made plans for his return, hopefully with Maryann, she fell pregnant gain so there was no hope of her being able to travel, even if James Baxter allowed it. (The baby boy, Harold Walker Smales was born in the July, a few weeks after Gideon arrived back at Hampton Park)

Gideon felt he couldn't delay his departure for much longer. His daughters were well established in their schools and John had returned to Cambridge and his book was published. Although very sad to leave Maryann and little Arthur behind he booked a passage on the 'Ruahine' leaving in the new year of 1868 and arriving in Wellington on 27th. May, thankfully arriving back at Hampton Park two days later.

Life back at Hampton Park, East Tamaki

With his usual enthusiasm, hard work and dedication Gideon immediately started to carry out all the improvements he had planned for the farm. He and James Stewart drew up plans to extend the house, building on several more rooms in preparation for the return of his daughters once they had finished their schooling and, hopefully, Maryann and the little boys. The girls were very keen to return to their home in New Zealand.

Gideon restocked his farm with the best animals he could afford, building up a milking herd, establishing heifers, steer, horses, sheep and pigs and chickens. His great love, however, was gardening and his plans for the gardens were incredibly ambitious. He employed a gardener, Old Campbell as he was known, who followed Gideon's directions and created a most beautiful garden with terraces, flower borders, a rockery, vegetable plots, box hedge gardens and orchards with every kind of fruit imaginable.

When Gideon was in England he was pleased to be re-connected with the English Wesleyan Conference especially as his severance from the New Zealand Wesleyans was complete and irrevocable. When he returned to New Zealand he continued his pastoral work and hoped to have a close association with the Anglicans in the Auckland area.

The Thames Gold Rush

On August 10th 1867 a prospector called William Hunt found gold in the bed of the Kuranui Stream and this was the start of gold mining in an area called Thames, in the region of Auckland. Gold had been found in other areas of New Zealand by ex-convicts and prospectors but it wasn't until 1867 that gold mining in that part of New Zealand became such a huge industry.

When news broke that gold had been found near Thames the sudden influx of prospectors increased the population of the small, mostly Maori, habitation by thousands with all the attending problems. Gideon was still building up his stock and organising his building projects but when news reached him of the wretched conditions the miners were living in he decided to go and see for himself. He was horrified by what he saw – men sleeping rough or if they were lucky in tents or basic huts made from raupo with no home comforts at all. He felt he had to do something practical to ease the situation and decided, at his own expense, to have constructed a building large enough to provide accommodation for some of the most deserving of the men.

The immense building, which Gideon called the Home Institute was completed within a few months. There were 100 small rooms or cubicles for sleeping and space for meetings and events which would help the community. To begin with everything went well and the local people were very pleased to have somewhere for social gatherings. As the prospectors began to make money they found their own places to live so the sleeping accommodation became less important and the hall was being used more for the social events rather than its initial purpose. Gideon was finding that the low rents that were being paid weren't enough to cover the cost of cleaning and general upkeep so what started as a great philanthropic gesture turned out to be a millstone around his neck. Eventually the building was converted into stables then sold on to a haulage company.

Gideon had more troubles and sadness in store so the failure of the Home Institute and loss of money meant little to him. He received a letter telling him that his little son, Harold, had died in April 1869 at just 10 months old so he never saw that little boy.

More worrying news about his son John came in a letter from the girls. John had graduated from Cambridge and had begun a post-graduate course in mathematics but owing to his deteriorating health he had to give up his studies just a month before his final exams and he returned to Esk Terrace. His sisters encouraged him to travel to France where the warm climate might help and he would escape the cold Yorkshire winter. With his health still deteriorating he returned to Whitby in the September where he was looked after by his sisters and Maryann, who was very poorly herself. John died, just 26 years old, on 16th September 1869 and was buried in Whitby cemetery. Just 9 days later poor Maryann died so the three girls had to arrange yet another funeral. It must have been the most devastating time for these young girls. A few months later more sadness followed. Gideon's little son, Arthur, aged just 4 died at Robins Hood's Bay near Whitby in the February of 1870 so he had lost all his boys and two wives in a matter of a few years.

Gideon received a letter of sympathy from Dr. Waddy and the senior lecturers at Cambridge saying that John, as a "premier young New Zealander" had achieved so much, gaining a first- class degree, and had he lived he would without a doubt have been accorded the prestigious Senior Wrangler award.

(Note – the Senior Wrangler is the top mathematics undergraduate at Cambridge, a position which has been described as the greatest intellectual achievement attainable in Britain.)

After these awful tragedies it must have been hard for Gideon to continue with life at Hampton Park but with his usual strength of character he carried on, not only with his farm work but also getting involved in the community. Now that the girls had finished their schooling it was nearing the time when he should be returning to England to bring them back home to Hampton Park. He had already put the properties he owned in Whitby on the market to raise funds to pay off loans and pay for his passage to England but before he had made any arrangements tragedy struck again. His eldest daughter Mary Anna (Polly) contracted TB and died at Hampton Park in August 1871 at just 27. Gideon and Polly's husband Samuel were devastated by the tragedy and it was a sad occasion when she was laid to rest at St. Johns Church where her little brother Gideon was buried. Samuel decided he could no longer stay on his farm in South Auckland and returned to England with his two small children.

Soon after Polly's death Gideon booked a passage for England going by the USA where he undertook a lecture tour partly to defray expenses. He arrived back in England in time to spend Christmas (1871) with his daughters in Esk Terrace and despite all the sadness it was a happy time for them all. Gideon again undertook another lecture tour, this time travelling around England for quite a few months, giving talks about life in New Zealand hopefully to encourage migration to the new country. During one of his visits to London he was introduced to a young woman, Elizabeth Tayler, and she made a great impression on him, so much so that he had hopes that she would become his third wife.

Elizabeth was an incredibly talented young woman. She spoke several languages particularly Italian and had worked in the household of a senior Italian Government official, Signor Muzzi, who eventually became postmaster general in Egypt. Elizabeth was governess to Signor Muzzi's children and for a while lived in Cairo with the family. Elizabeth was also a talented artist, singer, musician and gardener, the attributes that Gideon admired.

When Gideon met with Elizabeth again he proposed marriage and with her father's blessing she accepted his proposal and it was agreed that she would follow him out to New Zealand, travelling by sailing ship, and they would then be married in Auckland

Once Gideon completed his lecture tour he returned to Esk Terrace to make plans for their return home to Hampton Park. It was now late October and Gideon had booked a passage for himself and the girls on the Royal Mail steamer the 'Iberia'. They were travelling back to New Zealand via the USA and sailed from Liverpool on 7th. November 1872 bound for Boston. Before they left they bade a sad farewell to their friends and family but the girls were excited to be going home. Gideon hadn't mentioned to the girls that he had met Elizabeth although he did tell them that he might be getting married again.

The voyage to Boston took only two weeks which must have seemed quite speedy compared to the smaller sailing ships to which they were accustomed but the girls in particular had quite an uncomfortable time with often rough stormy weather.

The Great Fire of Boston

When the 'Iberia' sailed into Boston harbour on 20th. November they were met by an extraordinary sight. The whole of the commercial district had been destroyed by fire the previous week. The fire started on the evening of the 9th. in the basement of a commercial warehouse and rapidly spread to neighbouring buildings. Before the fire could be brought under control 12 hours later it had destroyed 776 buildings covering 65 acres. The commercial and financial districts were destroyed completely and the total cost was $73.5m overall. It was the most costly and destructive fire in the history of the USA.

There were several reasons why the fire spread so rapidly and why it took the fire department so long to get the fire under control.

It spread quickly because of the flammable wooden construction, particularly the wooden roofs, of the buildings and building regulations were not enforced. Gas supplies to the buildings and street lamps could not be shut off properly so the gas exploded and fed the flames. Flammable goods like wool, paper and textiles were stored in attics. Fire alarms were turned off to prevent false alarms; fire hydrants were not standardised; water pressure was low; a horse flu epidemic had affected the horses so the firemen had to pull the fire engines themselves. Amazingly there were only 13 fatalities but of course that was 13 too many.

Part of the devastation.

Gideon and the girls had a wonderful and exciting time while in America. They travelled through the Wild West to San Francisco on the First Transcontinental Railroad which was a more luxurious way to travel than they had been used to. They saw the Niagara Falls, Chicago, Native Indians, wolves and the Rocky Mountains, finally arriving in San Francisco after travelling more than 5000 miles. For the next part of the journey home Gideon and the girls boarded the 'Dakota,' a coal-fired steamer, on 6th December and after a stop at Hawaii they arrived back in New Zealand early in January 1873. After having been away for 10 years the girls were happy to be back home.

Back home

During Gideon's absence James Stewart had been completing the gardens and putting finishing touches to the church in anticipation of the return of Gideon and the girls and of course Gideon's future wife.

The girls were excited about the wedding and elaborate plans were made for the marriage ceremony and reception to which the community would be welcome.

The marriage took place on 14th March 1873 and was performed by the Anglican Bishop of Auckland, William Cowie with whom Gideon had formed a real friendship

Only a few weeks after the wedding Gideon received the sad news that his half- brother, Gideon 3rd, had died at the young age of 51 and this was followed a few months later by the death of his father 'Owd' Gideon in the January of 1873.

Gideon and his three daughters were delighted when Elizabeth told them that she was pregnant and on 4th. May twin girls were born but happiness turned to sadness when one of the little babies didn't survive. The baby who survived was named Elaine Elizabeth Tayler Smales.

Business was booming at that time in New Zealand with the economy thriving, immigration increasing and townships expanding although this was to the detriment of the Maori who were once again being treated as second class citizens and being squeezed out of their own environments.

Despite the many heartaches life, at this time, was also good for Gideon and his family. Gideon was now a successful farmer and land owner and was becoming more involved in the activities of the community around Auckland. The engagement of Sophie to a young farmer, Charles Overton, was welcome news and they were married at St. Johns in 1875. Elizabeth was pregnant again and soon after Sophie's wedding she gave birth to a baby boy, Herbert Michelle.

As fate seemed to decree both happiness and sadness were part of Gideon's life. Just when life seemed good with business booming and his involvement in the political scene giving him an interest tragedy struck. Susannah, (Rosie), just 26, was struck down with meningitis and died in May 1876 just three weeks before Elizabeth gave birth to a daughter, Alice Evelyn May.

The following year disaster of a different kind struck. After the boom years of the early 70.s the country began to go into recession with wool prices, particularly, falling rapidly followed by meat and other goods for the export trade. By 1878 the whole country was in a depression with unemployment and poverty and although Gideon and Elizabeth were better off than most they were beginning to feel the financial pressure. To raise funds Gideon was prepared to sell off part of his farm but land sales were slow or impossible. With deep regret he had to lay off some of his workers who had been with him for years, even James Stewart his right- hand man.

Unbelievably again happiness and sadness went hand in hand. In 1878 Elizabeth gave birth to their fifth child, a little girl, Adolie, then some months later Felicia was thrown from her horse and was very badly injured. She wasn't recovering well from her injuries and eventually died from TB in 1880.

Over the next four years Elizabeth gave birth to 3 more children, Beatrice Nares, Elfrida and a son Ambrose.

During those years Gideon spent much of his time writing, preaching, giving talks to local groups and keeping abreast of the politics of the time but Elizabeth and the children were the centre of his world and gave him a great deal of happiness.

In a cruel twist of fate Gideon's only surviving child, Sophie, from his family with Mary contracted meningitis and died aged just 34 in 1886. She was survived by her husband Charles and four small children and she was buried at Prebleton near Christchurch where Charles had his farm. All Gideon's 10 children from his marriages to Mary Bumby and Maryann Baxter died before him.

As age and those tragedies took its toll Elizbeth became the mainstay of the family and did her best to support Gideon. She was a great asset to the community as well – teaching, helping with the church activities, playing the church organ and giving singing recitals.

In his later years, after working tirelessly all his adult life for others and suffering so many personal tragedies it would seem only right that Gideon should have a few years peace and happiness but even in his old age he was beset by controversy. He was erroneously accused by "persons unknown" of illtreating a horse by putting it to work on his farm when it was suffering ill health. Gideon was humiliated at having to attend court but he defended himself robustly, denying the charge which he believed was a "put up job" for political or other reasons.

A government Committee had been set up some years previously to help those in financial or other difficulties through no fault of their own and claimants could apply for compensation. Gideon felt he had a good case as he had suffered financially during the recession because of the failure of creditors to pay him and the loss through theft of many of his farm animals. His main claim however was for compensation for the lecture tours he had undertaken during his visits to England which he had financed himself and for which he hadn't received any remuneration. The aim of his lectures and talks about New Zealand was to encourage immigrants to settle there so he felt that he had a creditable claim for compensation as immigration was so important for the developing country. The reason Gideon was so anxious to raise funds was to make life easier for Elizabeth and the children when he was no longer with them.

Although his health was failing Gideon was continuing to fight for what he thought was his right and what turned out to be his last visit to Auckland was to plead his case. He caught a chill on the journey home and became very sick. There was nothing the doctor could do as his heart was just worn out. He died peacefully on 5th. October 1894, aged 76.

The obituary, published by the New Zealand Herald on 6th October 1894 (see right) goes on to recount the life of this extraordinary man. Gideon Smales was not only extremely talented and hard-working but also a loving and devoted husband and father. He worked tirelessly in the Mission Field in Northland, New Zealand, making a tremendous difference to the lives of the local native population. The development of his farm, Hampton Park, and the building of the church there are legacies which continue to this day.

The legacy that Gideon Smales left to the New Zealand nation is undoubtedly the little church, St. Johns, which he built at his farm, Hampton Park, East Tamaki. The 60 seat chapel is regarded as a very important building as it is believed to be the oldest privately-built structure in New Zealand still being used on a regular basis for its original purpose (i.e. a church) It has been home to a small Anglican congregation for the last few decades and church services are held on alternate Sundays.

Gideon started the build in 1859 and it was completed by 1861. It was built from crushed shells from Howick beaches and basalt stone quarried from his own land and was the centre piece of his farm, Hampton Park. It was in this little 60 seat chapel that Gideon took church services after his retirement.

The church was damaged by fire in the 1960s but has been restored and a beautiful stained-glass window has been installed.

OBITUARY
Rev. Gideon Smales

DEATH OF THE REV. GIDEON SMALES.

WE regret to record the death of the Rev. Gideon Smales, of Hampton Park, East Tamaki, which took place yesterday morning at six o'clock. Only a few weeks have passed since Mr. Smales drove to the city as usual. He took cold on the journey home, and later it was found that the heart's action was very weak, not constitutional, but from old age. Dr. Bewes was called in, and everything that medical skill could suggest was done by the family to ease his pain and suffering, but the sytem had been too severely taxed, and he passed away peacefully as above.

Mr. Smales was born at Whitby, Yorkshire, on the 26th October, 1817, was accepted after examination by the English Wesleyan Methodist Conference as a Wesleyan minister, ordained at the Liverpool Conference of 1839, and left Bristol in September, 1839, in the Missionary brig Triton, 120 tons burden, with a large party of missionaries for New Zealand and other parts of the South Seas. The company included the Revs. Thomas Buddle, George Buttle, John Aldred, H. H. Turton, John Skevington, and Gideon Smales, so that we now have to record the death of the last of the Tritons.

Those very interesting episodes which have appeared in the columns of our Saturday Supplement during the past few months have given our readers some idea of the perils and dangers to which the

The Death of the Rev. Gideon Smales.
Gideon Smales died at his home, Hampton Park, on 5th October 1894. An excerpt from the obituary for Rev. Gideon Smales.

The 150th anniversary of the completion of the chapel was celebrated in 2011 with the Smales Trust, which looks after Hampton Park, organising all the activities. Members of the Smales family and other invited guests from around the world attended the celebration. The Smales Trust is aiming to promote the church and Hampton Park as a tourist attraction with walks through the park and guided tours of the old quarry, stone stable ruins and the remains of a beautiful sunken garden.

(Note – after Gideon's death Elizabeth and her children continued to live at Hampton Park and her son, Herbert, managed the farm.)

The New Zealand Company and the Wairau Affray

The New Zealand Company, restructured from The New Zealand Association, was founded in 1839 by Edward Wakefield, a London financier. His vision was to establish settlements in developing countries like New Zealand and Australia by buying or acquiring land, selling it in farm-size parcels to would- be immigrants and thus making a profit for the company and colonising these new countries.

Wakefield's plan was to sell the plots at a high enough price to ensure that the ordinary working classes couldn't afford to become landowners. They would, however, be encouraged to emigrate as there would be great opportunity for work on the farms of the rich landowners. This would ensure there was enough labour to work on the new farms. This, unfortunately, didn't always work out because many of the speculators simply bought land as an investment and had no intention of moving to New Zealand. Those absentee landlords didn't require labourers so many settlers who emigrated, expecting to find work, were left without employment.

The company agents, Wakefield's brother William being one of them, began to buy huge tracts of land from the Maori chiefs in the areas of Port Nicholson (later called Wellington) and Port Nelson, either side of Cook Strait. One problem arose which would have serious consequences was the fact that the Maori chiefs didn't understand English law or the legality of the conveyancing documents which they were asked to sign. Firstly, they couldn't read English so couldn't understand the documents. Secondly, they didn't understand the idea of change of ownership. To them the land deals simply meant "change of use" and they believed they still had ownership of their lands. Another point was that the Company were paying ridiculously low prices for the land, in some cases simply exchanging goods like guns and blankets, to make a good profit.

Concerns were being raised, not only by the missionaries in New Zealand but also by the British Government, that the Maori were being exploited over these land deals.

The British Crown had not considered colonising New Zealand at this time. The policy of the New Zealand Company was one of the driving forces behind the decision to instigate proceedings towards the colonisation of New Zealand. This led to the signing of the Waitangi Treaty in 1841 by the Maori Chiefs which led to New Zealand becoming a British colony. Once this had been achieved the Government decreed that any land sales by private companies, not approved by them, would be illegal. Realising this the NZ Company bought up land wholesale without regard to the Maori.

Edward Gibbon Wakefield.

More mistakes made

To attract settlers to New Zealand the Company publicised the new country as a land of opportunity, a perfect place with fertile farm land and a wonderful climate, a veritable paradise. They also depicted the native Maori as being keen to accept the "white man's ways" and would be eager to work for them on the farms. All negative reports were suppressed.

For many settlers what they experienced was not the paradise they had been promised. Some found that the land they had bought was useless, just scrub. Others found their plots didn't exist and many of course found there wasn't the opportunity of work. Many sold up, if they could find a buyer, and returned home disillusioned by the experience. Another mistake was that the company was selling plots of land which they hadn't yet bought and when the settlers arrived they found that it was questionable if they owned their plots. This led to trouble with the Maori who claimed they still owned their land.

The fact that the company was so keen to establish the settlements quickly meant that those new townships were disorganised and lacking in any governance. This resulted in trouble and discontent among the new settlers.

The first settlers arrived in 1840 followed by thousands during the next few years. The company commissioned special "migrant ships" to bring the settlers to New Zealand,

In mid-1843 the Company fell into financial difficulties, one reason being they were paying for free passage for the migrant workers. This was to ensure there were enough workers to satisfy the farmers, but as it turned out there were more workers than jobs available.

Colonel Wakefield resigned from the company in 1846 because of the financial difficulties incurred when land sales fell and poor decisions were made by the directors. They had applied to the British government for a loan which was granted but with strict conditions concerning their future dealings.

In 1850 the Company eventually admitted its mistakes and it was wound up eight years later. Colonel Wakefield, however, never admitted his failure and maintained that the new townships and settlements which were becoming successful were due to his vision of colonising new countries.

The Wairau Affray

The Wairau Affray of 1843 was the first serious war between the Maori and the Europeans but it certainly wasn't the last.

The New Zealand Company was realising that they hadn't acquired enough land for the numbers of settlers flooding into New Zealand. In 1843 they bought 70,000 acres, in an area near Port Nelson, from the widow of a Captain Blenkinsopp who had allegedly bought the land from the Maori. Although they anticipated trouble they went ahead with the deal.

The two chiefs of the Ngate Toa tribe were adamant they still owned the land and in revenge they destroyed some equipment and burned a few huts belonging to the Company. In retaliation Wakefield decided to arrest them and assembled a fighting force of about 50 special constables, some of whom were labourers who had no experience of handling weapons. On 17th. June 1843 Wakefield and his troops confronted the Maori and tried to handcuff the chiefs. In the scuffle a shot was fired, probably by accident by one of the untrained labourers, which instigated all out warfare. Four Maori were killed, including the wife of one of the chiefs, and several Europeans died too. When it became obvious that the Maori had the upper hand some of the Europeans fled and others surrendered. The Maori captured the prisoners but in revenge for the death of the chief's wife they proceeded to murder them by shooting or killing them with their tomahawks. In all 22 Europeans, including one of Wakefield's brothers, Arthur, and the Police Magistrate Thompson, were murdered.

The incident caused fear and anxiety among the settlers and it was the start of years of war and unrest between the two cultures.

In 1844 the new Governor, Captain Fitzroy, visited the Cook Strait settlement and exonerated the Maori, blaming the Europeans for their treatment of the native people and the 'taking' of their land. At that time the Europeans were furious with Fitzroy but now, in our more enlightened times, with hindsight, it is accepted that the Maori were treated disgracefully and the Europeans were in the wrong, thinking they could take what they wanted from the indigenous people without regard to their traditions, beliefs and the ownership of their own land.

The Wairau Affray was not the last conflict between the Maori and the settlers but after some years of unrest a relative peace prevailed.

The "*James*" sailing ship

John and Mary sailed to New Zealand on the "*James*", a barque sailing ship, which had been chartered to carry the party of Missionaries there.

A barque is a sailing ship with 3 or 4 masts with square sails, apart from the mizzen mast.

A barque
sailing ship

The "*James*" was built in 1832 from black birch, weighed 350 tons and belonged to the Port of London. The owners were Lidgett & Sons

When John and Mary arrived in Gravesend to embark Mary wrote in her journal that she was dismayed to see the ship which she thought was rather small for the large number of passengers and she felt very "dull" at the thought of travelling in such cramped conditions. Altogether there were 56 passengers on board, 21 being the missionaries and their families, the rest crew and other travellers. The captain's name was Todd and was apparently a very nice man but Mary does describe him as rather rough looking and hoped he improved on acquaintance. The "*James*" set sail on Thursday 20th September 1838 and the first port of call was the Isle of Wight where some passengers disembarked to have breakfast on the island and to buy a few items which had been forgotten. Even on this short journey the sea was stormy and some passengers suffered from sea sickness. One passenger decided not to continue and returned to dry land and the other passengers gave him a big cheer as he left the ship!

The 'Triton'

The Methodist Missionary Ship

On 7th May 1965 a lecture was given to the Wesley Historical Society, New Zealand, to commemorate the 125th anniversary of the arrival of the Methodist Missionary Ship '*Triton*' at Hokianga, Northland, New Zealand at 10pm on the 7th May 1840.

The lecture, given by Nora Buttle, told the story of the '*Triton*' and had been compiled from journals and diaries of some of the missionaries, notably the Revs Thomas Buddle, John Aldred and George Buttle, who had travelled on her on the voyage from England to New Zealand in 1839/40.

This chapter, telling the story of the voyage of the '*Triton*' from England to New Zealand in 1839/40, is based on the information in Nora Buttle's lecture. The quotes from the journals of some of the missionaries are also taken from the same source.

Preparation for the voyage

At the Annual Methodist Church Conference in August 1839 in Liverpool it was agreed that, to celebrate the centenary of British Methodism, a missionary ship should be purchased so that missionaries could be taken in greater numbers to distant places, particularly the South Seas Islands, Tasmania and New Zealand. At the previous year's conference it had already been discussed and the finance committee had allotted funds towards the purchase of the ship.

Below is an excerpt from the report:-

For Missionary Polynesian Ship purchase £3000

For outfit of same – stores, insurance, incidentals............£3000

Before the 1839 conference concluded a committee of men, knowledgeable in seamanship, was appointed to find a suitable ship and make all the necessary arrangements to ensure she was ready to sail in mid -September of that same year.

The missionaries who were to travel to the South Sea Islands, including New Zealand, had already been appointed and included were Rev T Buddle, Rev Turton, Rev Gideon Smales and Rev George Buttle. The Rev and Mrs Archill and their family of 6 children were also travelling but just as far as Cape Town.

The '*Triton*'

The ship that the committee purchased was the '*Triton*', a 119 ton brigantine, which had been built in Bristol in 1837 and had been used in the sugar trade in the West Indies. Although it was considered by some to be rather small for the purpose the committee thought it a suitable choice for sailing around the islands in the South Seas.

A huge amount of work needed to be carried out on the '*Triton*' to make her comfortable for the passengers and safe for the long voyage as well as being well stocked with food and other essentials. Alterations had to be made to the interior of the ship to create sufficient cabin room for the missionaries and the crew and a lower deck was laid to allow space for cabins and cargo.

"The dining salon measures 8ft by 18ft and here arrangements are made for two of the single missionaries to sleep on what were called 'sofas' in those days. Every berth has a patent

light and from three skylights there can be a constant current of air. Arrangements are made for a shower bath, and near that room for two drip-stones and one of Robins patent filters."
(Excerpt from one of the journals)

A round house was built on deck to give some protection from the elements for the 1st and 2nd officers while on duty. Work on the ship also included stripping, felting and coopering plus three new sets of sails with several bolts of spare canvas. Three boats, two of which were new, plus 'Trengrouses life apparatus' donated by a friend in Cornwall, were installed for the safety of the passengers in the event of an incident while at sea.

(Note – Henry Trengrouse, born in Helston, Cornwall, in 1772 trained as a cabinet maker. In 1807 he witnessed the sinking of HMS Anson off the Loe Bar when the Captain and over 100 sailors lost their lives. He was so affected by this tragedy that he devoted the rest of his life to devising life-saving equipment. He invented the rocket-powered rescue system for ships in distress which he named the 'Bosun's chair'. It ultimately saved over 20,000 lives over the years and is now known as the 'Breeches Boy'. He also designed a life jacket.

The British Government did not recognise his incredible achievements and gave him a mere £50 for an order of 20 of the systems then, having studied the design, went on to manufacture the systems themselves. Trengrouse, however, was recognised by the Russians as many of their sailors had been saved in the Baltic using his system and they invited him to Russia to develop his invention - an offer he declined. The Tzar of Russia gave him a diamond ring and the Society of Arts awarded him their silver medal and a grant of £30.

Henry Trengrouse died in 1854, a poor man financially, and is buried in the churchyard in Helston. Although, disgracefully, the British Government did not recognise the value of his work he is still honoured in Cornwall and many countries around the world and the system is found around our shores.

The main thoroughfare in Helston is Turngrouse Way and an example of his life-saving apparatus is on display in the Folk Museum in Helston.)

The Crew

The Captain and crew were carefully selected not only for their seamanship but for their sympathy towards and allegiance to the Methodist Faith.

The Captain, John Beatty, was an experienced sailor and a gentleman who treated his passengers and crew in a very courteous manner. His wife, who accompanied him on the voyage with their two infant children, was the daughter of a Methodist minister so they were in tune with the missionaries.

The chief officer was Mr. Buck, a onetime Methodist class leader and a devout Christian. He had served as captain on previous voyages and was Captain Beatty's righthand man.

There were 10 crew most of them of the Methodist Faith which was important to the Mission Society.

The craftsmen who were carrying out the alterations to and improvements on the ship must have had to work incredibly quickly as the date for departure was set for 12th September, just 5 to 6 weeks since the ship had been purchased.

Departure

The passengers had arrived in Bristol ready to sail on the 12th and were looked after by the Bristol Methodists until the ship sailed. Conditions, unfortunately, were not right for sailing on that date so departure was delayed with the missionaries hearing the words *tomorrow will be the day*. The following day the weather was fair but again the message *tomorrow will be the day* was heard. The delay that time was due to the Captain's wife having just given birth to her second child. The sailing had to be postponed till the 14th so that she and her day-old baby and her other infant could board the 'Triton' ready to accompany her husband on the long voyage.

On the day of departure there was great excitement with the missionaries and their friends arriving in hackney carriages and other vehicles at the docks. With relief they saw the *'blue peter'* flying at the 'Triton's' masthead – the sign that the ship was ready to sail.

At 10am precisely a tug towed the 'Triton' slowly down the channel with invited friends and mission secretaries on board who were given a tour of the ship. They were shown all the alterations that had been made, all the stores – enough for up to 4 years long voyage – and the gifts for the various missions and other places that would be visited en route.

There wasn't room for all the goods and what couldn't be fitted in would be taken by other ships to Hobart, Tasmania, to be picked up by the 'Triton' on arrival in the South Seas. During the day there was a special service of baptism for the new baby and at 4pm the visitors and friends prepared to leave the ship after a service of prayers and hymns.

After the excitement of the departure the passengers' thoughts turned to what lay ahead and the friends and family left behind. A feeling of gloom descended on the company as they assembled on the deck and unfortunately this was replaced by an unpleasant feeling of sea sickness and they quickly found their way to their cabins searching out the various cures they had brought with them - essence of ginger, essence of peppermint, Seidlitz Powder and brandy – but nothing helped! The 'Triton' unfortunately in rough seas had a rolling twisting motion which resulted in many passengers suffering sickness many times even in the first few days.

(Note – Seidlitz Powder a digestion regulator, was manufactured from the the 19th century. The ingredients were tartaric acid, potassium sodium tartarate and sodium bicarbonate.

The name derives from the village of Sedlec in the Czech Republic, previously Bohemia, a spa famous for its Sedlitz waters, although there is no connection between the two.)

After 4 days due to the stormy weather they had made little headway and were no further than Milford Haven. The Captain decided to seek shelter there much to the relief of the passengers who enjoyed being on '*terra firma*' again. Since repairs had to be carried out and additional stores collected it was October 1st before they were underway once more.

Within two days of leaving Milford Haven the ship ran into stormy weather again and the Captain realised they were in a very difficult predicament as they were reduced to only one lifeboat should the need arise and in rough seas that one wouldn't be able carry its quota of ten in such heavy seas. Of the three boats ordered for the '*Triton*' one had been left behind to make room for cargo and another had been lost in the storm. One diarist wrote "*so in very truth their only hope is the good Providence of God, throughout the whole voyage their faith in God is very real and never wavers*".

On October 14th they were in sight of Madeira Island and two days later were anchored in one of the Bays and the passengers, much to their relief, were able to go ashore and were welcomed by Rev. Mr. Edwards, a (Dissenting) Clergyman from Nottingham, who was on the Island for a health-restoring holiday. He had seen their ensign bearing the text "*Glory to God*" so realised the passengers were of the Church. They found Madeira to be a quiet, peaceful and beautiful place and were more fortunate than Mary as the ship remained there for a full week whereas Mary and her companions only saw the Island from their ship which had passed the island a year earlier.

The Harbour Master and the Health Officer made a very perfunctory inspection of the ship, not even going on board, and all the passengers simply lined up on the side while the inspection was made by the officers from their boat.

The passengers had a very pleasant time on Madeira visiting the Cathedral, churches, monasteries and even a nunnery although the missionaries were not very impressed by the way the church services were conducted "*it appears a mockery of Divine Worship*" wrote one diarist.

Six of the brethren took a ride through the town on horseback each horse being accompanied by its Portuguese owner who kept hold of the horse's tail.

"*And I thought* (wrote Brother Turton) *this was a test of strength between man and steed so I let go and we galloped from one street to another, uphill and down, until the poor man exhausted began bawling most vociferously*".

Evidently the owners used to keep hold of their horses' tails in this way to stop strangers stealing them and could run for miles at a normal trot but not at a gallop."

On 23rd October the ship set sail again just in time for the celebration and special services on the 25th October which is the date named for Centenary Celebration throughout the Methodist Community. Even the crew were remembered on the special day and were given extra rations – goose and plum pudding.

The typical daily routine on the 'Triton' was as follows ;-

8am – the Bell for Morning Prayers called the passengers to the first service of the day.

Breakfast followed and consisted of "*strong burnt coffee, sad hot rolls, unpeeled potatoes, sea biscuits and cold salt beef*".

After breakfast, if the weather was fair, a stroll round the deck - in small groups because of lack of space – was enjoyed then the men went to their cabins to study and the women did needlework, read and wrote letters.

12 o'clock was lunch time – called '*tiffin time*' – and at 1.30pm the children and some of the women had dinner, the others at 3pm. It was difficult to have everyone in the small dining room at the same time.

Being a missionary ship there were regular church services and when practicable they were held on deck with the binnacle for the pulpit and chicken coops for the pews but this had its drawbacks as the poultry clucked and quacked throughout and pecked the legs of whoever was sitting on their coop.

A large part of the day was spent by the missionaries in study and religious services which must have been difficult at times with 8 young children, one a tiny baby, on board and not much space for them to run around although the missionaries did plan amusements for them. For the adults amusement it was decided to issue a weekly paper, '*The Triton Observer*', to appear on Friday evenings but it only ran to three issues because a letter written to the paper about the "*propriety of dress on board the ship*" upset and hurt some of the passengers so it was decided it would be wise to cease publication!

As the '*Triton*' approached the Tropics they experienced a tornado which nearly engulfed the little ship. "*Without a moment's warning a tremendous squall came up, accompanied by torrents of rain, with heavens as black as midnight; the sea rose instantly; and then in the space of an hour wind and rain ceased and sky bright and clear.*"

The ship crossed the equator about noon on Wednesday 20th November but, being a missionary ship, the usual celebrations were very low key with just a few buckets of sea water being thrown over some unsuspecting souls as they strolled on deck.

Since approaching the Tropics the passengers had suffered storms, a tornado and suffocating heat and then on 22nd December another possible threat of a different kind occurred. A Brig had been seen on the previous evening by the watch and it appeared to be manoeuvring in a suspicious manner. Then on the morning of the 22nd the Brig approached in a direct line – it appeared to be flying the Portuguese Ensign and Pennant – and a boat was lowered with crew and a "*tall moustachioed vulgar sort of a gentleman dressed in shabby fine uniform and whose countenance evinced no kind of sympathy whatever with the better feelings of our nature.*"

All the men on board the '*Triton*', including the steward, the cook and the cabin boy, lined up on the side of the ship, from stem to stern, to make it appear the ship was well manned in case of trouble. The foreigner, accompanied by a seaman as interpreter, came on board and through his interpreter stated he was the Lieutenant of the Portuguese Brig-of-War carrying troops to South America. This was obviously false as the Brig had no gun ports, was sailing in the wrong direction for South America, and there appeared to be only three crew on board. The 'Lieutenant' spoke in a foreign language, presumably Portuguese, but when he jotted down the information given him by Captain Beatty, he wrote in the 'Queen's English' as one of the missionaries noticed as he peeped over the man's shoulder. The diarist wrote of the event "*It was the opinion of us all that for once we had come across a Pirate and we were left in a state of suspense as to the probable course which would be adopted after the officer reached his ship. And we ourselves were in a state of the greatest helplessness; all our arms and ammunition being stowed away under the cargo; yet we felt the disposition to defend ourselves.*"

It was with relief that they saw the Brig sail away in the direction of Africa and fortunately they didn't encounter it again. The next day was spent clearing a way to the arsenal, oiling rusty muskets, getting out and adjusting long and short boarding pikes and making other preparations for defence in case the Brig returned.

As Christmas approached the brethren prepared themselves for the Christmas and New Year's celebrations – the Love Feast on Christmas Eve, two services on Christmas Day, a solemn Watch Night service on New Year's Eve when they all would renew their consecration vows. As it happened the ship was becalmed on Christmas Day 1200 miles off the coast of South Africa.

South Africa

On the morning of 3rd January 1840, the welcome cry of "*land in sight*" was heard by the passengers and they were all happy to hear the ship was close to Saldanha Bay about 60 miles NW of Cape Town and they were hoping to be on dry land the next day. The contrary weather, however, prevented that and strong gales blew the ship further from the coast. In desperation, and because the passengers were suffering seasickness again, the Captain decided to risk an attempt to reach the shelter of a cove in the Bay and was fortunately successful. They sheltered there for some days.

Because there had been so many delays during the voyage they were running short of provisions and water so the Captain and some of the men used the small boat to go ashore hoping to find a village or some dwellings to buy some food. They came across some fishermen who told them there were only a few Dutch farmers in the neighbourhood and a British Resident to look after them and there were no extra stores to sell. The following day some of the men again went ashore to search for stores, again unsuccessful, and by the time they arrived back on the beach night had fallen so they couldn't see the anchorage and had to sleep on the shore using their upturned boat for shelter. The next day, fortunately, they found a Dutch farmer who could supply them with their needs and they returned to their boat with a few sheep, a calf and a good supply of fresh water, all carried slung over their shoulders. They were greeted with great shouts of welcome when they rowed back to the 'Triton'.

After many attempts to get the ship out of Saldanha Bay the Captain eventually anchored the 'Triton' in Table Bay and the Methodist Missionary, the Rev. Hodson, came aboard to welcome the party to Cape Town. He had been looking and waiting for the 'Triton' for two months so was very relieved to see her anchored in the Bay. When the passengers came ashore they found themselves in a completely different world.

"The jetty was crowded with goods of all kinds and men of all colours, countries and characters. What with the shipping and landing of merchandise, the rumbling of wagons, the cracking of whips, the shouts of the boatmen and all the confusion of tongues – African, Eastern and Continental – the noise and jargon was complete and should have served as a scene to relieve the stupor and monotony of a long tropical voyage like ours,, but for the depressing thought that most of its actors were without God and without hope in the world."

The Rev. and Mrs. Archbell and their six children and the Rev. and Mrs. Appleyard were leaving the 'Triton' to start their missionary work in Cape Town which would make the next part of the voyage a little less crowded and a lot quieter much to the relief of the brethren. The remaining missionaries were able to relax and enjoy their stay in Cape Town which they described as *"a beautiful and delightful place with a large population made up of many nationalities – English, Dutch, Malayans, Chinese and others – with well- built stone buildings, plenty of shops, warehouses, public buildings, and no lack of churches and chapels, but religion they felt to be at a low ebb."*

The Captain took the opportunity to dismiss two of his crew who had been ill mannered, foul mouthed and unruly during the voyage so he spent his time looking for replacements.

During the day, around noon, when the weather was very hot and the South Easter sprang up blowing dust everywhere, the shops closed and the locals retired to rest until around 3pm but the missionaries spent the time reading, studying, listening to music and writing letters home.

The missionaries, as they explored Cape Town, did become aware of and were appalled by the discrimination towards the black-skinned people who were despised and disregarded. When, however, they visited an Infant School belonging to the Methodist Society they found boys and girls of different races being cared for and treated equally as 'one' which they thought must have an influence in days to come of a lessening of the idea of a superior caste.

One day they came upon a small cottage hospital some miles out of the town where small pox patients were being cared for. This disease had just been found in Cape Town and to control the spread of infection the surrounding area of the hospital was cordoned off and the brethren were not allowed to approach, although they would have liked to visit and pray with the sick.

One morning, very early at 5am, the brethren set out to visit the Barrack Yards where more than 700 slaves, men, women and children, were congregated. Having been freed from captured ships these poor people, without knowing what was happening, were to be apprenticed to different farmers and tradesmen in the neighbourhood for some years. The brethren were appalled at this heart-rending sight of these poor dejected people being treated in this way.

The brethren met up with another Methodist Missionary who had emigrated from Barnsley in Yorkshire quite recently and were delighted to meet a fellow Yorkshireman especially the Rev. Buttle who hailed from Snaith near Barnsley. They also enjoyed their visits to the vineyards where the grapes were cheap and plentiful and as many as they wanted could be bought for just a small sum.

From Cape Town to Hobart Town.

After a stay of three weeks in Cape Town the passengers boarded the 'Triton' on 13th February and they set sail for Hobart, Tasmania.

The voyage from now on was a little more peaceful and roomy as the Archwells plus their six children and the Appleyards remained in Cape Town. Captain Beatty had replaced the crew members who had been unruly and ill-mannered and the new crew turned out to be excellent, especially a native from the West Indies who looked after the passengers well.

When the 'Triton' ran into stormy weather the passengers again suffered from seasickness and the West Indian, promoted to steward, proved to be a great help and "*every evening he would continue to keep as much fire in the galley as would boil a large saucepan of gruel, for it was a correct opinion of his that a farinaceous diet was far the best in cases of sickness, and this he would bring from cabin to cabin.*"

After that one storm the weather was good for the rest of the voyage to Tasmania and the missionaries settled down to studying, reading and planning their mission duties. They also found great pleasure in gazing at the stars and "*on clear evenings we were blessed with a view of the heavens and found objects enough in the rising stars and constellations to attract our deepest study*" and they thought the Southern Hemisphere much superior to that of the Northern.

Early on the morning of Saturday 4th April the passengers heard the man on watch shout out "*Hullo there, land in sight*" and very quickly all were on deck looking excitedly towards Van Diemens Land as it was named by the diarist (in 1856 the name of the Island was changed to Tasmania). There they were to meet the Rev. John Waterhouse who was the General Superintendent of the Methodist South Sea Mission and they fully expected to be on shore in a few hours.

As often happened, however, the wind dropped when the '*Triton*' was just 8 miles off shore and was becalmed for a further three days which was frustrating for passengers and crew alike. On April 7th a pilot came on board and the '*Triton*' sailed up the Derwent River and before evening was safely anchored in Sullivan's Cove. The missionaries were greeted by the Rev. Waterhouse and the Rev Nathaniel Turner who had worked in the mission fields of New Zealand and the Pacific islands for many years.

"His copper-coloured care-worn sexagenarian-looking visage easily bespoke the amount of service he had done in the mission fields and to us he appeared to be a man from another world, the Apostle of Polynesia."

As in Cape Town the brethren were looked after and entertained by the local missionaries and they felt quite at home in Hobart as it seemed very English with its broad streets, brick buildings, churches and chapels, warehouses and offices and private houses and the beautiful river Derwent as it wound in and out of the picturesque scenery.

The Hobart Methodists appeared prosperous owning valuable property with chapels, a Day School and another chapel under construction. The missionaries joined in with the services with enthusiasm describing the "Tea Meetings" as being "*just like a bit of England once more.*"

The missionaries stayed for 14 days in Hobart during which time they celebrated Easter with special services and a special 'Gathering for Sunday School children' with tea and buns.

While in Hobart the crew of the '*Triton*' stocked up with stores – flour, tea, coffee, sugar, salt etc. – and picked up the boxes and packages which had been left for them by other ships.

The original plan was for the '*Triton*' to sail to Sydney before heading for New Zealand but as there had been so many delays the Captain and the Rev. Waterhouse decided to sail directly to New Zealand much to the consternation of the passengers who had been counting on doing last minute shopping in Sydney. The decision wasn't finalised until 10pm so from then until midnight the passengers were "*sailing about the town in all directions, rousing up the shopkeepers, ransacking their stores, and from the sharp looks which some of them kept over our movements, subjecting ourselves to the awkward suspicion of being well-dressed London sharpers!*"

Journeys End

The following day the passengers with the Rev. Waterhouse and his family boarded the 'Triton', all their luggage stowed away, farewells bade to their hosts and at 4pm, when all the guests had returned to shore, the 'Triton' once again set sail, heading for New Zealand.

After suffering more storms by Sunday 2nd May the 'Triton' was within 100 miles of the New Zealand coast but another severe storm, when even the Captain thought that the little ship would founder, prevented them making progress and it was a further 4 days before they were close to Hokianga Harbour. Again the weather was against them and they were becalmed but within sight of their destination.

Deciding to wait no longer the missionaries and crew lowered the boats and they "*vied with each other in towing the 'Triton' a distance of 10 miles so as to be ready to cross the much-dreaded bar with the sea breeze at the time of high water.*" They waited anxiously for a pilot to arrive to guide them over the bar into the safety of the harbour but as no one appeared Mr. Waterhouse decided to go up river by boat with a crew to find a pilot as by now the 'Triton' was nearly on the rocks and in danger of sinking. They met up with a Captain Young who, realising the danger the ship was in, organised additional boats and crew and by 10pm the 'Triton' was safely anchored in the river near the Newark Mission Station about 5 miles from Mangungu. The chief pilot was evidently "ill in bed" or to be nearer the truth was "dead drunk".

The 'Triton' remained at Hokianga for the next two weeks and the missionaries and Rev. Waterhouse visited Mangungu meeting with Mr. Hobbs and John Bumby and others who had waited so long for the arrival of the 'Triton'.

The Rev John Bumby wrote "*Mangungu 12th April 1840. It is impossible to convey upon paper anything like a correct idea of the 'raru raru', as the natives say, i.e. the bustle and confusion that prevail at present at Mangungu. I cannot allow the opportunity to pass without informing you after weeks and months of distressing suspense and anxiety our eyes were refreshed and our hearts gladdened by the arrival of the long-expected 'Triton' with her 'precious cargo' all alive and well.*"

The 'Triton' set sail again on 19th May heading for Fiji and Tonga where some of the missionaries were bound. John Bumby joined the ship to sail south to Kawhia which lies on the coast just south of Auckland where he had missionary work to do.

The 'Triton'

For the next six years the 'Triton' continued to sail round the New Zealand coast and the Pacific Islands taking missionaries to and from their respective mission stations and collecting and delivering stores but after these years of essential service she was replaced by a newer ship, the 'John Wesley', a larger and better equipped vessel.

The '*Triton*' was sold by auction in April 1847 for £900 but she would be remembered as a tough little ship which weathered many storms under the skilled hands of Captain Beatty and Chief Officer Buck.

The Mangungu Mission House

Th Mangungu Mission House, "Horeke's historic gem", is situated on the slopes of Horeke overlooking the picturesque Hokianga Harbour. It was built in 1838/1839 to replace the original Mission House which had been burned to the ground in August 1838.

Both Mary and John record in their journals that they had heard of the disaster and Mary, particularly, was concerned as to what they would find when they arrived and where they would stay

An entry in "The Life of John Hewgill Bumby" by Alfred Barrett, published in 1859, records how the fire started. He writes

"In August 1838 a serious and distressing disaster befell the Mission settlement in Mangungu. Mr. Turner had retired to rest as usual on Saturday night, the 18th. after leaving a log on the fire of the room which they usually kept by day, in order that he might, if needful, be able to provide for the wants of his invalid partner..............................About 2 o'clock he was awakened by a roaring noise like that of a fire"

Mr. Turner, because of the smoke and heat, had to escape and call for help. Fortunately, being a Saturday, there were many natives who had arrived for the Sunday service and they all helped to save Mrs. Turner and the children but couldn't douse the flames and many of Mr. Turner's books and papers and the personal belongings of the family were lost. The Mission House was destroyed completely.

When Mary and John arrived at Mangungu they found that the Mission House had been completely rebuilt.

The newly built house which replaced the original was designed and built by the Rev. John Hobbs, a missionary who had been apprenticed to his father who was a carpenter and joiner and agricultural implement maker in England. The Rev. Hobbs was indeed a very talented man as he not only designed the elegant Georgian building but he oversaw the construction.

One very important event which took place at the Mangungu Mission House was the signing of the Waitangi Treaty in 1840 between the British Government and the Maori chiefs.

In 1856 the Mission House was being used less as the Maori population was declining in that part of the country so the Rev. Hobbs decided to move to Auckland and he took with him not only his family but also the Mission House, leaving the Mission unmanned. The

The Mangungu a Mission a House, circa 1840, taken from an old drawing.

Drawing of the table at which
the Treaty was signed.

house was taken to pieces and shipped to Onehinga where he was living and then later it served as a parsonage.

In 1972 the Historic Place Trust of New Zealand realised the importance of the building and the house was bought and brought back to Mangungu where it was rebuilt on the original site and is maintained by the Trust and is now open as a Museum.

The Museum has many articles on display connected to the time that Mary and John Bumby and other Missionaries were there.

The actual table at which the Waitangi Treaty was signed is in the Museum as well as portraits of Missionaries and their books and papers. There are everyday utensils which would be used in the kitchen and other everyday objects.

There has recently been a Cycle Trail established to help increase visitor numbers. The Museum is very important to the community and to the heritage of New Zealand.

Mangungu Mission House.

The Waitangi Treaty

The Treaty of Waitangi is an agreement made between the representatives of the British Crown and the Maori chiefs of New Zealand. It was signed by over 500 Maori chiefs in 1840.

At that time the population of New Zealand was chiefly the indigenous Maori and about 2000 Europeans. New Zealand was being used as an overflow for the penal colony in New South Wales, Australia and as well as the convicts and ex-convicts there were gold miners, prospectors and land developers. The British Government did not have any ambition to colonize New Zealand at that time and there was little law enforcement in place to deal with any crime or unrest.

There was one official, James Busby, who had been appointed British Resident, an official title, in 1833 and he was tasked with trying to protect the 'well-disposed settlers and traders', prevent 'outrages' against the Maori and to apprehend any escaped prisoners.

The attitude of the British Government changed when it was becoming apparent that the Maori were being exploited by land-grab developers, especially the New Zealand Land Development Company based in England. To protect the Maori and their customs and land, also to make sure they had a fair deal on any land sale, the British Government decided to lay claim to New Zealand and, with the agreement of the Maori chiefs, make it a colony of the British Empire. It was also in the British Government's self -interest. Another determining factor in this decision was that the French Government was also very keen to colonise New Zealand so the British wanted to step in first.

In 1840 Naval Captain William Hobson was sent out as Lieutenant Governor and it was his task to negotiate for the sovereignty of New Zealand (or those parts that would agree) and establish a British Colony.

Hobson arrived at the Bay of Islands on 29th January, 1840 and with his secretary's help and advice from George Gipps, Governor of New South Wales, he made some notes for a treaty. James Busby tidied them up and made some additions. The Treaty was translated into Maori by the Rev. Henry Williams, a missionary, with the help of his son, Edward.

The chiefs were invited to Waitangi on 5th Feb. to discuss the Treaty and on 6th Feb. 1840, about 50 chiefs agreed to sign the document. The Treaty was taken around the country for more chiefs to sign and on the 12th Feb. the largest signing took place at the Mangungu Mission House. Because John Bumby was a long way away, preaching in a different part of the country at that time, Mary Bumby acted as hostess to the large number of people who attended the signing. Altogether over 500 chiefs signed, including a small number of women. Many signed using a sign or symbol as few Maori could read or write at that time.

The meaning of some words, unfortunately, in the Maori translation differed from the English version so the Maori had different expectations of the terms. For some words there was no equivalent in Maori and one important example was the use of the word 'sovereignty' which had been translated as 'governance'. This, to the Maori, meant that they expected full authority over their customs and treasures and rights. There were several copies of the Treaty and each was witnessed by a European representative. Nine of these copies have survived and are held in the Archive Museum of New Zealand.

Commemoration

Every year on 6th Feb. the signing of the Waitangi Treaty is marked and since 1974 has been a public holiday. Waitangi Day is recognized as New Zealand's National day but even today there are still disagreements and tension associated with the interpretation of the treaty.

The Waitangi Treaty House

The Treaty House at Waitangi is the former home of James Busby, the British Resident at the time of the signing of the Treaty in 1840. It was also the site of the Declaration of Independence of New Zealand in 1835. In 1932 the house and grounds were bought by the Lieutenant Governor and donated to the nation and it became a National Reserve in 1934. It is the site of the annual Waitangi Day Celebration.

Mary Bumby's Bees

It is thought that honeybees (*Apis mellifera*) evolved many millions of years ago in Africa and gradually moved into Europe as the climate became warmer so it is a European species.

The European honeybee, however, is now found throughout the world. In countries like America, Australia and New Zealand honeybees were not an indigenous species but were taken there by travellers.

Honeybees first arrived in Virginia on the east coast of America in 1622, on the sailing ship the '*Discovery*' as part of a consignment of goods like seeds, fruit trees, rabbits and pigeons. Bees also arrived into other parts of America in the 1800's and 1900's taken by travellers so the bees were eventually widespread throughout North America.

In 1809 honeybees were taken to Alaska but there is no record as to whether they survived,

In 1821 honeybees arrived in Australia on the sailing ship the "*Mary*" and the following year bees were brought there on the "*Isabella*". In 1824 honeybees were taken to Tasmania on the sailing ship the "*Phoenix*" and in 1839 Mary Bumby, from Thirsk, Yorkshire, was the first to take honeybees to New Zealand.

How could the bees survive such long voyages but survive they did. The beekeepers transporting them must have known how to care for them and prepare them for the voyage.

The bees at that time were kept in skeps so were easy to transport. Some sort of carrying box would be needed to keep the skeps safe while being loaded onto the ship and during the voyage.

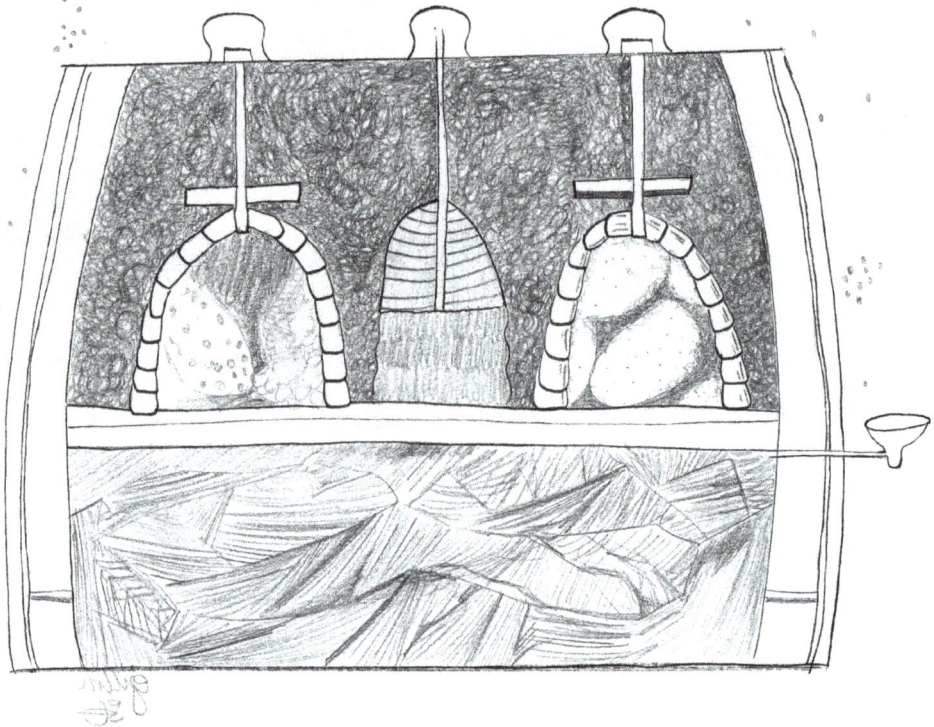

Illustration of a carrying box showing insulation beneath the skeps, ventilation pipes and a small funnel presumably for feeding syrup to the bees.

It would have been quite feasible for Mary, during the voyage, to have the skeps taken up on deck during suitable calm weather. This would allow the bees to have a "cleansing flight" thus keeping their nest clean. This is what bees do naturally and during our winters they will come out for a flight when the weather is mild.

The bees wouldn't fly far from their skep and would take short orientation flights so they would know the position of their skep entrance and they would return after a few minutes.

Bees dislike flying over water as it seems to disorientate them as there are no landmarks to guide them so on board the ship they would keep within range of their skeps.

If they had plenty of stores, were kept dry and had good ventilation and had the opportunity for the occasional flight they could certainly survive several months at sea.

Although Mary didn't mention her bees in her journal they did survive the long months at sea. When the 'James' arrived at Mangungu the bees were placed in the Mission church-yard as being a safe place for them. The native Maori, not having heard of or seen honey-bees, were very wary of them and kept away. On an early drawing only one skep is seen so maybe only one colony survived – or perhaps the second one was in a different area of the

Record of Mary's bees from her original stock – 1843 – 44

Hives names – King Henry viii, Edward, Marianne, Samuel, Henry

Date 1843	wt. of honey	swarms	hive
December 27th	46lb honey	4	King Henry viii
March 28th	8lb	4	Edward
July 18th	28lb	0	Marianne
1844			
October 2nd	2, 4, 8lb	4	Samuel
1845	Henry died out		

Honey was taken in both summer and winter

And the strongest colony was King Henry viii

Record of the weight of honey taken and the
number of swarms from Mary's hives.

churchyard. There is no doubt, however, that Mary's honeybees were the first to arrive in New Zealand and those bees would have come from Thirsk in Yorkshire and would have been the local native strain of the British black bee.

(Note – In a letter dated 1988 to a Miss Bumby from Mr W R Bielby in New Zealand he wrote;-

"At Rowena I found a few black bees foraging in a bottle brush bush. This was unusual as nearly all New Zealand honeybees are yellow-banded Italian bees. Could these bees be descendants of the original colony?")

Mr Bielby is correct in stating that the New Zealand bees nowadays are very distinctive with yellow/gold banded bodies and a quiet nature. This is due to the fact that the light-coloured very productive Italian bees and queens were imported in great numbers into New Zealand in the 20th century when commercial beekeeping expanded and became a very important business for the country.

The local flora and pleasant climate would suit Mary's bees and they would do very well and would spread by swarming into the locality.

In a New Zealand publication called The Beemaster one contributor tells of how he had access to some apiary notes made in 1843-45 by a 'relation' or friend of Mary Bumby. Presumably with a growing family and helping her husband in his mission work Mary would have little time for beekeeping so a friend or 'relation' would have taken over the work.

By 1843-45 there were apparently five colonies and the notes record the amount of honey taken and the number of swarms from each hive.

The hives were given names with No, 1 being called King Henry V111 and the others were Edward, Marianne, Samuel and Henry. The notes record that Henry died out in October 1844.

King Henry V111 was the most productive producing 46¼ lbs of honey and four swarms in 1844.

The manuka plant grows widely in the North Island of New Zealand so Mary's bees would undoubtedly visit the plants and produce manuka honey. In recent years manuka honey has become very much sought after because of its healing properties and its taste. Mary and her brother and friends would therefore be the first to eat New Zealand manuka honey.

Map of the Route of the 'James'

Margate, September 21st

Madeira, October 8th

Canary Islands, October 11th

Tropic of Cancer, October 13th

Off Trinidad, November 27th

Tropic of Capricorn, November 29th

South Africa – Cape Town, December 19th

Isle de St. Paul – Indian Ocean, January 12th

Storm Bay, River Derwent, February 1st

Hobart Town, February 2nd

Left Hobart, March 8th

Hokianga River, March 18th

Arrived at Mangungu, March 19th

Tropic of Cancer

The Equator

Tropic of Capricorn

Map of Place Names

North Island, New Zealand

Waimate

Bay of Islands

Mangungu

Waitangi

Hokiannga
Harbour

Horeke

Kaipora
Harbour

East Tamaki

Auckland

Onehinga

Porirua

Port Nelson

Port Nicholson
(Wellington)

Wairau Valley

Cook Strait

David F Bumby's Ancestors

Stephen Bumby, born in Appletreewick, was baptised in his local parish church at Burnsall, North Yorkshire on 12th June 1659. His father was Anthony Bumby who would have been born around the 1630's. Stephen must have grown up and married in the area as he too had a son called Anthony baptised in Burnsall on 19th December 1680.

The registers of St Mary's Parish Church in Thirsk, North Yorkshire reveal that an Anthony Bumby, a blacksmith, had a son Humphrey baptised 1714. Although it cannot be proved it seems likely that the Burnsall Anthony and the Anthony recorded in Thirsk are the same person. Burnsall and Thirsk are about 30 miles apart but why he may have moved remains a mystery.

Anthony is the earliest member of the Bumby family mentioned in the Thirsk Parish records when his son Humphrey was baptised on 28th December 1714. He died in 1724 shortly before the birth of his fourth daughter Frances. The Thirsk records include over 120 references to Bumbys between 1714 and 1837 and the family gradually spread out from Thirsk into the surrounding parishes. There are still many family members with the surname Bumby resident in the Thirsk area to this day.

Humphrey Bumby, born in 1714, followed his father and became a blacksmith as did two of his sons, Abraham and John. Another son, Anthony, was a shoemaker. Sadly, three of Humphrey's children died in infancy and he died in 1787.

Anthony, the shoemaker, was the father of John, a butcher, who married Mary Hewgill in the village of Hawnby near Thirsk in 1803. Their two children, John Hewgill Bumby and

Mary Bumby are the subject of this book. They are somewhat distant relatives of mine, 2nd cousins, 3 times removed.

John Bumby, the blacksmith, son of Humphrey was my 3x great grandfather. He was born in 1743/44 and married Ann Ward in 1774. Their family consisted of eight children. Of the four boys three became blacksmiths and one of the three, Abraham, died on the island of St Vincent in the West Indies in 1823. John and Ann's youngest son, another John, was born 4th February 1794 and is my 2x great grandfather.

This John married twice. His first wife, Mary Pierson bore him three children but she died aged 37. John subsequently married Margaret Weighill some six years later and had a further five children. He apparently broke the blacksmith mould and became the innkeeper of the White Swan Inn in Thirsk Market Place. He died aged 80 in Middlesbrough.

John and Margaret's second son was James Bumby who was born 3rd February 1838, my great grandfather. James spent most of his life working on the railways in various roles and, at times, was out of work. In 1861 he was described as a railway clerk in Leeds. Later he became station master at Starbeck near Harrogate. According to the 1871 Census he was a "Railway Clerk – Out Employ", in 1881 a quarryman and finally back to being a railway clerk in 1891.

James married Jane Shepley Lister in Leeds in April 1861 and, it seems, was the first of my branch of the family to move away from Thirsk since 1714. James and Jane had eight children, five girls and three boys one of whom, Francis Lister Bumby was my grandfather. James died in 1906

Francis was a Post Office telegraph engineer and was born in Starbeck. One of the projects he worked on was the installation of telegraph lines to the racehorse stables in Middleham, North Yorkshire where he met and subsequently married my grandmother, Xarifa Sarah Auton. His specialist skills resulted in him being enlisted into the Royal Engineers during the First World War even though he was 48 years of age, well over the normal age for conscription. He assisted in laying and repairing communication lines in the trenches. Both Francis and Zarifa died long before I was born but I do recall speaking to Francis's older brother Charles on the telephone in 1954 when he celebrated his 90th birthday.

Over the last couple of hundred years the surname Bumby has spread around the world although the main concentration is still in North Yorkshire. The name can be found in such countries as the USA, Canada, South Africa and Australia.

David F Bumby

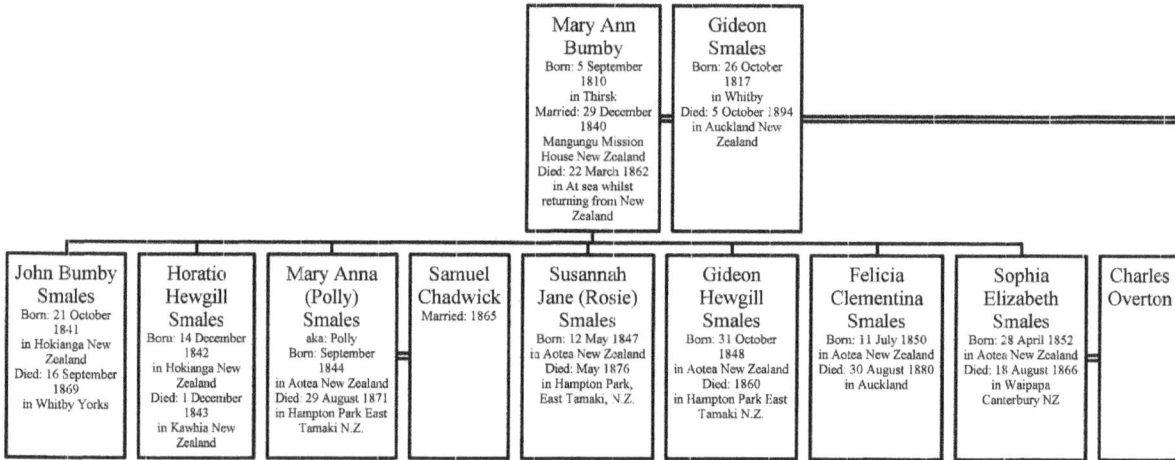

Mary Ann Bumby
Born: 5 September 1810
in Thirsk
Married: 29 December 1840
Mangungu Mission House New Zealand
Died: 22 March 1862
in At sea whilst returning from New Zealand

Gideon Smales
Born: 26 October 1817
in Whitby
Died: 5 October 1894
in Auckland New Zealand

John Bumby Smales
Born: 21 October 1841
in Hokianga New Zealand
Died: 16 September 1869
in Whitby Yorks

Horatio Hewgill Smales
Born: 14 December 1842
in Hokianga New Zealand
Died: 1 December 1843
in Kawhia New Zealand

Mary Anna (Polly) Smales
aka: Polly
Born: September 1844
in Aotea New Zealand
Died: 29 August 1871
in Hampton Park East Tamaki N.Z.

Samuel Chadwick
Married: 1865

Susannah Jane (Rosie) Smales
Born: 12 May 1847
in Aotea New Zealand
Died: May 1876
in Hampton Park, East Tamaki, N.Z.

Gideon Hewgill Smales
Born: 31 October 1848
in Aotea New Zealand
Died: 1860
in Hampton Park East Tamaki N.Z.

Felicia Clementina Smales
Born: 11 July 1850
in Aotea New Zealand
Died: 30 August 1880
in Auckland

Sophia Elizabeth Smales
Born: 28 April 1852
in Aotea New Zealand
Died: 18 August 1866
in Waipapa Canterbury NZ

Charles Overton

Descendants of Gideon Smales

Maryann Baxter		Elizabeth Taylor
Born: in Whitby Married: 30 November 1864 Whitby Died: September 1869 in Whitby		Married: 14 March 1873 St John's Church,Hampton Park, East Tamaki

Arthur Baxter Smales	James Smales	Harold Walker Smales	Eline Elizabeth Smales	Twin Of Elizabeth Smales	Herbert Michelle Smales	Alice Evelyn May Smales	Adolie Smales	Beatrice Nares Smales
Born: 2 September 1865 in Whitby	Born: April 1867 in Whitby Died: April 1867 in Whitby	Born: July 1868 in Whitby Died: April 1869 in Whitby	Born: 4 May 1874 in Hampton Park, East Tamaki	Born: 4 May 1874 in Hampton Park, East Tamaki Died: 4 May 1874 in Hampton Park, East Tamaki	Born: 1875 in Hampton Park, East Tamaki	Born: June 1876 in Hampton Park, East Tamaki	Born: 1878 in Hampton Park, East Tamaki	Born: in Hampton Park, East Tamaki

Elfrida Smales	Ambrose Smales
Born: in Hampton Park, East Tamaki	Born: in Hampton Park, East Tamaki

References

Smales' Trail by John Steele

The Life of the Rev. John Hewgill Bumby by Alfred Barrett

The Voyage of the Triton by Nora Buttle

Mary Bumby's journal

David Bumby's archive re Mary Bumby and her family

Mangungu Mission House Museum

New Zealand history website

'Papers Past' website

'World Cat' website

STUFF.co.nz website

Manuka – the Biography of an Extraordinary Honey by Cliff Van Eaton

Photo credits

www.ingramcontent.com/pod-product-compliance
Lightning Source LLC
Chambersburg PA
CBHW050257090426
42734CB00022B/3482